Orestes N. Stavroudis

The Mathematics of Geometrical and Physical Optics

Orestes N. Stavroudis

The Mathematics of Geometrical and Physical Optics

The k-function and its Ramifications

WILEY-VCH Verlag GmbH & Co. KGaA

The Author
Prof. Dr. Orestes N. Stavroudis
Centro de Investigaciones en Optica,
León, Guanajuato, Mexico
ostavro@cio.mx

Cover picture
Persistence of Vision Raytracer Version 3.5
Sample File
Author: Christopher J. Huff

Library of Congress Card No.:
applied for

British Library Cataloguing-in-Publication Data
A catalogue record for this book is available from the British Library.

Bibliographic information published by Die Deutsche Bibliothek
Die Deutsche Bibliothek lists this publication in the Deutsche Nationalbibliografie; detailed bibliographic data is available in the Internet at <http://dnb.ddb.de>.

Printed in the Federal Republic of Germany

Typesetting Steingraeber Satztechnik GmbH, Ladenburg

Printing Strauss GmbH, Mörlenbach

Binding Schäffer GmbH, Grünstadt

ISBN-13: 978-3-527-40448-3
ISBN-10: 3-527-40448-1

Acknowledgements

Many hands make the work light. It is my pleasure to acknowledge and thank those whose efforts made this work possible. These are (in alphabetic order): Maximiliano Avendaño-Alejo, Isidro Cornejo, Lark London, Christopher Stavroudis, Dorle Stavroudis, and many former students whose helpful comments and snide remarks were of immensurable value. Thanks are also due to my project editor, Ulrike Werner, who skillfully and tactfully, led me in the proper direction.

So I pass from a task, which has filled the greater part of many years of my life, which has broadened in my view as they passed, and which has suffered interruptions that threatened to end it before its completion. Many of its defects are known to me; after it has gone from me, others will become apparent. Nevertheless, my hope is that my work will ease the labour of those who, coming after me, may desire to possess a systematic account of this branch of pure mathematics.

A. R. Forsyth
Trinity College, Cambridge
October, 1906.

Introduction

This work is about geometrical optics though it shall extend into some fundamental areas of physical optics as well. It makes heavy use of several branches of mathematics which, perhaps, the reader will find disturbingly unfamiliar. These I will describe with some care but with only lip service to mathematical rigor and vigor.

Keep in mind that geometrical optics is a peculiar science. Its fundamental artifacts are rays, which do not exist, and wavefronts, which indeed do exist but are not directly observable. A third item is the caustic, a surface in image space which is certainly observable, defined variously as the envelope of an array of rays associated with some point object, the locus of the principal centers of wavefront curvatures, or as the locus of points where the differential element of area of a wavefront vanishes. Of course, these wavefronts must be in a wavefront train generated by a lens and associated with some fixed object point.

The peculiarities of geometrical optics go even further. Rays, which do not exist, are trajectories of corpuscles, which also do not exist. These trajectories, according to the principle of Fermat, are those paths over which the time of transit of a corpuscle, passing from one point to another, is either a maximum or a minimum.

Yet it works. Geometrical optics, anachronistic as it is, remains the basis for modern optical design, the highly successful engineering application built on the sandiest foundation imaginable. There is hardly one area of modern science in which instruments are used whose design depends ultimately on Fermat's postulate on the intrinsic laziness of mother nature.

In what follows I shall use a method best described as axiomatic, the axiom being Fermat's principle. This we must modify, however. Since point-to-point transit times can be maxima as well as minima we must use, in the language of the Calculus of Variations, *extrema* (singular: *extremum*) as our criterion in applying Fermat's principle.

Indeed the interpretation of the principle of Fermat in terms of the language of the variational calculus will lead us to ray paths in *inhomogeneous* media; media in which the refractive index

The Mathematics of Geometrical and Physical Optics: The k-function and its Ramifications. O.N. Stavroudis

ISBN: 3-527-40448-1

is a continuous function of position. These ray paths will be expressed in the form of a system of ordinary differential equations that can be applied to any specified media.

These ray paths are then subject to analysis using the techniques of the differential geometry of space curves. Using these differential equations for a ray path we can deduce its shape and its relationship to the refracting medium itself. From these results we can determine, quickly and easily, the nature of rays in, say, Maxwell's fish eye.

From here we pass on to the Hilbert integral, developed originally for dealing with the problem of the variable end point in the Calculus of Variations. This very rich theory leads us to a number of very important deductions in geometrical optics; conditions for the existence of wavefronts, Snell's law, The Hamilton-Jacobi equations (though both Hamilton and Jacobi preceded Hilbert by as much as a half century), the eikonal equation, among others. In this context the theorem of Malus becomes trivial. From this context Herzberger recognized the importance of the *normal congruence* or the *orthotomic system* of rays.

With the concept of the wavefront in hand we proceed to the differential geometry of surfaces and to partial differential equations of the first order. One such is the eikonal equation, mentioned above, obtained from the Hamilton-Jacobi equation, for which we find a general solution descriptive of any wavefront train in a *homogeneous optical medium*; one with a constant refractive index.

In terms of the differential geometry of surfaces we can find, for the general wavefront train, wavefront principal directions and curvatures. This leads to the important concept of the caustic, that surface that is the locus of the principal centers of wavefront curvature. In the caustic resides all of the monochromatic aberrations associated with a wavefront train and, ultimately, with the lens and object point that give rise to it. The structure of this caustic describes completely the *image errors*: *spherical aberration, coma* and *astigmatism*. Its location in space indicates the *field errors*; *distortion* and *field curvature*.

Along the way we look at *generalized ray tracing*, more properly, a generalization of the *Coddington equations*, that determines the principal directions and principal curvatures at any point on a wavefront through which a traced ray passes.

This we apply to the prolate spheroid, a rotationally symmetric ellipsoid generated by rotating an ellipse about its major axis. This leads to a reflecting optical system, consisting of two confocal spheroids, that I have called the *modern schiefspiegler*.

We also look at Herzberger's fundamental optical invariant and his diapoint theory and apply it to the representation of wavefronts obtained form the solution of the eikonal equation. This leads to a hierarchial system of aberrations.

The canon that I have described here, based on Fermat's principle, omits many important items. Outstanding among these is *paraxial theory* and *paraxial ray tracing*. Although is it is of tremendous practical importance, it is based on an approximation that, in my opinion, does not belong here.

A far more fundamental omission is Gaussian optics, in particularly, its model as developed by Maxwell. He began with certain assumptions about perfect lenses from which he represented perfect image formation by a fractional-linear transformation. Upon assuming that his perfect lens is rotationally symmetric, he was able to derive its cardinal points; the foci, the nodal points and the principal points.

Other omissions are the *Seidel aberrations* and their higher order extensions. These are from a solution of the eikonal equation in the form of a power series that has never been shown to converge.

Huygens' principle is omitted. It is clearly independent of any corpuscular concepts and is based on wavefront propagation as the envelope of spherical wavelets, which also do not exist, centered on a previous position of the wavefront. It also leads to Snell's law. It was for many centuries the main competitor to Fermat's corpuscules.

But nowadays the photon incorporates the best of both the corpuscule and the wavelet, a compromise that has resulted in a far more useful theory with applications far beyond the dreams of Fermat and Huygens.

Contents

The Mathematics of Geometrical and Physical Optics: The k-function and its Ramifications. O.N. Stavroudis

ISBN: 3-527-40448-1

Part I

Preliminaries

The Mathematics of Geometrical and Physical Optics: The k-function and its Ramifications. O.N. Stavroudis

ISBN: 3-527-40448-1

1 Fermat's Principle and the Variational Calculus

In the seventeenth century light was believed to be a flow of *corpuscules*, 'little bodies'; their trajectories were called *rays*. Pierre de Fermat asserted that Nature was intrinsically lazy and that those corpuscules 'chose' a trajectory that made their time of transit from point to point a minimum. We refer to this anthropomorphism as *Fermat's Principle*. It was a successful hypothesis. With it, Fermat was able to derive the law of refraction, *Snell's law*, in an economical and precise way.[1]

The connection between Optics and the variational calculus came some years after Fermat when the Swiss mathematician Jacob Bernoulli proposed a problem, the *brachistochrone*, and offered a prize for its solution. Consider a rigid wire connecting a pair of points, fixed in space, on which a bead slides under the force of gravity but without friction. The problem was to find that shape of the wire for which the time of transit of the bead, from one point to the other, was a minimum.[2]

The connection between geometrical optics and Fermat's principle is clear. Jacob's solution was to calculate the vertical force on the bead, taking into account the constraint imposed by the rigid wire. He related this force to an index of refraction function that depended on the height of the bead on the wire. He partitioned the space between the initial and terminal points into horizontal lamina each having a constant refractive index that was determined by its height. Then he could use Snell's law to trace a ray down from the initial point, resulting in a polygonal ray path that approximated the desired solution. As the number of lamina increased and as each thickness approached zero, the polygonal figure approached a continuous curve which was the desired shape of the rigid wire. This curve turned out to be an arc of a cycloid.[3]

Jacob Bernoulli was very pleased with his solution, so much so that he awarded to himself the prize that he had offered, and disregarded the efforts of his brother Jean, who also solved the brachistochrone problem, from an entirely different point of view.

Jean made use of the newly discovered differential calculus and the fact that the first derivative of a function vanishes at its maximum or minimum value. He expressed the time of transit of the bead from the initial point to its terminal point as a an integral of the reciprocal of its velocity. The first derivative of this integral must vanish at a minimum and he obtained conditions that the solution curve must satisfy. Subsequently Leonard Euler extended Jean's

---

[1] Sabra 1967, Chapter V. An account of the history and background of Fermat's principle.

[2] Bliss 1925, pp. 65–72. Caratheodory 1989, pp. 235–236 uses the Hamiltonian which we will encounter in Chapter 3. Woodhouse 1964, Chapters I and II provides a more detailed historical account. Courant & Robbins 1996, pp. 381–384. In Smith 1959, pp. 644–655 there is an English translation of Bernoulli's original paper and announcement.

[3] Bliss 1946, Chapter VI. Jean's use of the calculus in generating the variational calculus.

The Mathematics of Geometrical and Physical Optics: The k-function and its Ramifications. O.N. Stavroudis

ISBN: 3-527-40448-1

method to more general problems and obtained differential equations for their solution. Jean's method can rightfully be called the beginning of modern Calculus of Variations.[4]

It is natural to refer to a solution of a variational problem as an *extremal arc* or more simply as an *extremal*. We will interpret the principle of Fermat in terms of the language of the variational calculus and apply modern mathematics to that basic axiom of geometrical optics and develop it as far as we can.

1.1 Rays in Inhomogeneous Media

We have seen that the basic assumption of geometrical optics is Fermat's principle: A ray path that connects two points in any medium is that path for which the time of transit is an extremum. To be more explicit, out of the totality of all possible paths connecting the two points, A and B, a ray is that unique path for which the time of transit is either a maximum or a minimum. Of course if A and B are conjugates, if B is a perfect image of A, then the ray path is not unique; every ray passing through A must also pass through B.

The time of transit between two points, A and B, is given by the equation

$$T = \int_A^B dt = \int_A^B \frac{ds}{v} = \int_A^B \frac{nds}{c}, \tag{1.1}$$

where c is the velocity of light *in vacuo*, v its velocity in the medium through which it propagates and n the refractive index of that medium. The arc length along the ray or trajectory is s. The optical medium is said to be *homogeneous* if n is constant; it is *inhomogeneous* but *isotropic* if n is a function of position. It is *anisotropic* if the refractive index of the medium depends on the ray's direction.

The convention most used is to drop c from the equations and to use the *optical path length* I, instead of the *time of transit* T, as the variational integral. Thus

$$I = \int_A^B nds. \tag{1.2}$$

In what follows we take the medium to be inhomogeneous so that the refractive index is a function of position $n = n(x,\ y,\ z)$. A possible path connecting A and B is given parametrically by the three coordinate functions $x(t),\ y(t),\ z(t)$ where the choice of the parameter t is entirely arbitrary. If A has the coordinates $(a_1,\ a_2,\ a_3)$ and B, $(b_1,\ b_2, b_3)$ then it must be that

$$\begin{aligned} x(t_0) &= a_1, \quad y(t_0) = a_2, \quad z(t_0) = a_3, \\ x(t_1) &= b_1, \quad y(t_1) = b_2, \quad z(t_1) = b_3, \end{aligned} \tag{1.3}$$

[4] Bliss 1946, Chapter I. Bolza 1961, Chapter 1. Clegg 1968, Chapter 3.

so that

$$I(A,\, B) = \int_{t_0}^{t_1} n(x,\, y,\, z) ds = \int_{t_0}^{t_1} n(x,\, y,\, z) \frac{ds}{dt} dt, \tag{1.4}$$

where the Pythagorean theorem gives us

$$\frac{ds}{dt} = s_t = \sqrt{x_t^2 + y_t^2 + z_t^2}. \tag{1.5}$$

Here, the subscript $(_t)$ denotes differentiation with respect to the parameter t. This subscript notation for both ordinary and partial differentiation will be used extensively in what follows.

In these terms then the problem is to find that curve, given by $x(t),\ y(t),\ z(t)$, for which $I(A,\, B)$ is an extremum.

1.2 The Calculus of Variations

This problem is a special case of a more general problem that belongs to that body of mathematics known as the Calculus of Variations. That more general problem is to find the curve in space, given by $y(x),\ z(x)$ for which the integral

$$I = \int_a^b f\big(x,\, y(x),\, z(x),\, y_x(x),\, z_x(x)\big) dx, \tag{1.6}$$

is an extremum. The function f is always known since it is determined by the nature of the problem; for example, in Eq. 1.4, f is equal to $n(x,\, y,\, z) ds/dt$.

Here we need to find expressions for $y(x)$ and $z(x)$ that make Eq. 1.6 an extremum. First assume that $\overline{y}(x)$ and $\overline{z}(x)$ represent a solution, a curve for which Eq. 1.6 is an extremum. In addition let $\eta(x),\ \zeta(x)$ be any two functions, sufficiently differentiable, such that

$$\begin{aligned} \eta(a) &= \eta(b) = 0, \\ \zeta(a) &= \zeta(b) = 0. \end{aligned} \tag{1.7}$$

Now form a one-parameter family of curves given by

$$y(x) = \overline{y}(x) + h\,\eta(x), \qquad z(x) = \overline{z}(x) + h\,\zeta(x), \tag{1.8}$$

where h is the parameter. By virtue of Eq. 1.7 these curves all pass through the end points of the integral; when the parameter h is *zero* we have, by definition, the solution curve. We replace $y(x)$ and $z(x)$ in the variational integral, Eq. 1.6, by using Eq. 1.8 to get

$$\begin{aligned} I(h) &= \int_a^b f\Big(x,\, \overline{y}(x) + h\,\eta(x),\, \overline{z}(x) + h\,\zeta(x), \\ &\qquad \overline{y}_x(x) + h\,\eta_x(x),\, \overline{z}_x(x) + h\,\zeta_x(x)\Big) dx. \end{aligned} \tag{1.9}$$

Because of our construction, if $h = 0$ then I is at an extremum value and, for that value of h, dI/dh must vanish. We calculate this derivative, set it equal to zero and get, from Eq. 1.9

$$\left.\frac{dI}{dh}\right|_{h=0} = \int_a^b \left\{ \frac{\partial f}{\partial y}\eta + \frac{\partial f}{\partial z}\zeta + \frac{\partial f}{\partial y_x}\eta_x + \frac{\partial f}{\partial z_x}\zeta_x \right\} dx = 0. \tag{1.10}$$

Apart from the properties given in Eq. 1.7, the functions $\eta(x)$ and $\zeta(x)$ are entirely arbitrary, a fact that will be important later.

We expand Eq. 1.10 using integration by parts. Recall that,

$$\int_a^b u\, dv = u\, v\Big|_a^b - \int_a^b v\, du,$$

so that

$$\int_a^b \frac{\partial f}{\partial y}\eta\, dx = \left[\eta \int_a^x \frac{\partial f}{\partial y} dx\right]_a^b - \int_a^b \left[\int_a^x \frac{\partial f}{\partial y} dx\right] \eta_x dx. \tag{1.11}$$

Since η vanishes at a and b, the first term vanishes. In exactly the same way we get

$$\int_a^b \frac{\partial f}{\partial z}\zeta\, dx = -\int_a^b \left[\int_a^x \frac{\partial f}{\partial z} dx\right] \zeta_x dx. \tag{1.12}$$

Substituting Eqs. 1.11 and 1.12 into Eq. 1.10 results in

$$\int_a^b \left\{ \left[\frac{\partial f}{\partial y_x} - \int_a^x \frac{\partial f}{\partial y} dx\right] \eta_x + \left[\frac{\partial f}{\partial z_x} - \int_a^x \frac{\partial f}{\partial z} dx\right] \zeta_x \right\} dx = 0. \tag{1.13}$$

Note that if the quantities in brackets are constant then the integral vanishes and the condition is satisfied.

This condition is also sufficient. Recall that our choice of the functions η and ζ is completely arbitrary. For the integral to vanish for all possible choices of these functions then the coefficients of their derivatives in Eq. 1.13 must be constant.[5] We conclude that

$$\begin{cases} \dfrac{\partial f}{\partial y_x} - \displaystyle\int_a^x \frac{\partial f}{\partial y} dx = \text{constant} \\[2ex] \dfrac{\partial f}{\partial z_x} - \displaystyle\int_a^x \frac{\partial f}{\partial z} dx = \text{constant}. \end{cases} \tag{1.14}$$

[5] Bliss 1946, pp. 10–11 calls this the *Fundamental Lemma of the Calculus of Variations*. I believe that the proof given here is simpler.

If f possesses second derivatives we get the *Euler equations*

$$\begin{cases} \dfrac{d}{dx}\dfrac{\partial f}{\partial y_x} &= \frac{\partial f}{\partial y} \\ \dfrac{d}{dx}\dfrac{\partial f}{\partial z_x} &= \frac{\partial f}{\partial z}, \end{cases} \tag{1.15}$$

a pair of simultaneous *ordinary* differential equations. Recall that f describes the nature of the particular problem and therefore must be known. The solution is an extremal arc that connects the fixed initial and terminal points. Each pair of these end points provide boundary conditions that define a solution. The aggregate of all such solutions to Eq. 1.15 is called a *field of extremals.*

We will need yet another relationship. The total derivative of f with respect to x is

$$\begin{aligned} \frac{df}{dx} &= \frac{\partial f}{\partial x} + \frac{\partial f}{\partial y}y_x + \frac{\partial f}{\partial z}z_x + \frac{\partial f}{\partial y_x}y_{xx} + \frac{\partial f}{\partial z_x}z_{xx} \\ &= \frac{\partial f}{\partial x} + \left[\frac{d}{dx}\frac{\partial f}{\partial y_x}\right]y_x + \left[\frac{d}{dx}\frac{\partial f}{\partial z_x}\right]z_x + \frac{\partial f}{\partial y_x}y_{xx} + \frac{\partial f}{\partial z_x}z_{xx} \\ &= \frac{\partial f}{\partial x} + \frac{d}{dx}\left[y_x\frac{\partial f}{\partial y_x} + z_x\frac{\partial f}{\partial z_x}\right], \end{aligned} \tag{1.16}$$

in which we use the Euler equations, Eq. 1.15 to get

$$\frac{\partial f}{\partial x} = \frac{d}{dx}\left[f - y_x\frac{\partial f}{\partial y_x} - z_x\frac{\partial f}{\partial z_x}\right]. \tag{1.17}$$

1.3 The Parametric Representation

The problem can also be expressed in parametric form.[6] We represent the arcs connecting the two end points, A and B, by the coordinate functions $x(t)$, $y(t)$ and $z(t)$ of the arbitrary parameter t. It must be that when $t = a$, all possible arcs must pass through A and when $t = b$ they must all pass through B. With this proviso the variational integral becomes

$$I = \int_a^b f\big(x(t),\, y(t),\, z(t),\, x_t(t),\, y_t(t),\, z_t(t)\big)dt. \tag{1.18}$$

It is important, indeed vital, to understand that the parameter t must be applied *uniformly* to all of these possible paths connecting A to B. The *choice* of the parameter t is unimportant and can be anything convenient.

However the choice of t cannot effect the statement of this variational problem and therefore any transformation of t must leave the structure of Eq. 1.18 completely unchanged. To show this [7] we use the *reductio ad absurdum* argument; we assume the contrary and demonstrate a contradiction. First assume that f does indeed depend explicitly on t so that it takes the form

$$f = f\big(t,\, x(t),\, y(t),\, z(t),\, x_t(t),\, y_t(t),\, z_t(t)\big).$$

[6] Bliss 1946, Chapter V. Bolza 1961, Chapter IV. Clegg 1968, Chapter 7.
[7] Bliss 1946 Chapter V. Theorem 41.1.

If we apply a linear transformation to t, say, $t \to \tau + h$, then the variational integrand becomes,

$$f\big(\tau + h, x(\tau + h), y(\tau + h), z(\tau + h), x_\tau(\tau + h), y_\tau(\tau + h), z_\tau(\tau + h)\big)d\tau.$$

Since the differential of τ cannot contain the constant h the transformed variational integrand does not have the same structure as the original version. This contradiction proves that f cannot depend on t explicitly.

We can take this a little further. Suppose the transform involves a factor as in, say, $t \to h\tau$ so that the variational integrand takes the form,

$$f\big(x(h\tau),\, y(h\tau),\, z(h\tau),\, x_\tau(h\tau)/h,\, y_\tau(h\tau)/h,\, z_\tau(h\tau)/h\big)h d\tau.$$

Compare this expression with the integrand in Eq. 1.18. For this expression to have the same structure as the original variational integrand, f must be a homogeneous function[8] of $x_t,\ y_t,\ z_t$. That is to say,

$$f\big(x,\, y,\, z, \lambda x_t,\, \lambda y_t,\, \lambda z_t\big) = \lambda f\big(x,\, y,\, z, x_t,\, y_t,\, z_t\big).$$

Taking the derivative of this expression with respect to λ, then setting $\lambda = 1$, yields

$$f = x_t \frac{\partial f}{\partial x_t} + y_t \frac{\partial f}{\partial y_t} + z_t \frac{\partial f}{\partial z_t}, \tag{1.19}$$

showing that f must indeed be a homogeneous function in $(x_t,\ y_t,\ z_t)$.

To summarize these results: A variational problem in terms of a parameter t cannot depend on t explicitly; moreover f must be a homogeneous function in $x_t,\ y_t$ and z_t.

In Chapter 5, in which we look at partial differential equations, we will show that a general solution of Eq. 1.19 is obtainable and that the solution is indeed homogeneous; the condition is therefore sufficient as well as necessary. Observe that Eq. 1.19 is the analog of Eq. 1.17 which, in this parametric case, is trivial.

Again we assume a solution, $\overline{x}(t),\ \overline{y}(t),\ \overline{z}(t)$ and choose arbitrary functions $\xi(t), \eta(t),\ \zeta(t)$ that vanish when $t = a$ and when $t = b$, then form the variational integral

$$I(h) = \int_a^b f(\bar{x} + h\xi, \bar{y} + h\eta\ , \bar{z} + h\zeta\ , \bar{x}_t + h\xi_t,\ \bar{y}_t + h\eta_t,\ \bar{z}_t + h\zeta_t)dt. \tag{1.20}$$

We go through the same steps as before and get

$$\begin{cases} \dfrac{d}{dt}\dfrac{\partial f}{\partial x_t} = \dfrac{\partial f}{\partial x} \\[2ex] \dfrac{d}{dt}\dfrac{\partial f}{\partial y_t} = \dfrac{\partial f}{\partial y} \\[2ex] \dfrac{d}{dt}\dfrac{\partial f}{\partial z_t} = \dfrac{\partial f}{\partial z}, \end{cases} \tag{1.21}$$

the Euler equations for the parametric case.

[8] Rektorys 1969, pp. 454–455.

1.4 The Vector Notation

The vector notation simplifies greatly the results obtained for the parametric case. Suppose we have some differentiable function $f(x,\ y,\ z)$. Then its total differential is,

$$df = \frac{\partial f}{\partial x}dx + \frac{\partial f}{\partial y}dy + \frac{\partial f}{\partial z}dz.$$

(Of course f can have any number of independent variables but for our purposes *three* is exactly right.) This can be written as a scalar product of two vectors,

$$df = \left(\frac{\partial f}{\partial x},\ \frac{\partial f}{\partial x},\ \frac{\partial f}{\partial x}\right) \cdot (dx,\ dy,\ dz)\,.$$

The left vector we identify as the *gradient* of f

$$\nabla f = \left(\frac{\partial f}{\partial x},\ \frac{\partial f}{\partial y},\ \frac{\partial f}{\partial z}\right). \tag{1.22}$$

If we let $\mathbf{V} = (x,\ y,\ z)$ then the total derivative in vector form is

$$df = \nabla f \cdot d\mathbf{V}. \tag{1.23}$$

When cast in vector form the results of the last section assume a much more compact form. We first define the vector function of the parameter t,

$$\mathbf{P}(t) = \big(x(t),\ y(t),\ z(t)\big);$$

its derivative with respect to t must then be,

$$\mathbf{P}_t(t) = \big(x_t(t),\ y_t(t),\ z_t(t)\big),$$

and the variational integral defined in Eq. 1.18 becomes

$$I = \int f(\mathbf{P},\ \mathbf{P}_t)dt. \tag{1.24}$$

Moreover, as was shown in the last section, f must not depend on t explicitly and it must also be homogeneous in $\mathbf{P}_t$.

Next, define two vector gradients according to Eq. 1.22

$$\begin{aligned} \nabla f &= \left(\frac{\partial f}{\partial x},\ \frac{\partial f}{\partial y},\ \frac{\partial f}{\partial z}\right), \\ \nabla_t f &= \left(\frac{\partial f}{\partial x_t},\ \frac{\partial f}{\partial y_t},\ \frac{\partial f}{\partial z_t}\right). \end{aligned} \tag{1.25}$$

Applying these to Eq. 1.21 we get the vector form of the Euler equations

$$\frac{d}{dt}\nabla_t f = \nabla f. \tag{1.26}$$

Because f is homogeneous in $\mathbf{P}_t$ it must be that $f = \nabla_t f \cdot \mathbf{P}_t$, this from Eq. 1.19.

In conclusion one might say that the application of the Calculus of Variations consists of two parts; stating the *question* and getting its *answer*. The *question* part is finding the f-function appropriate to the application. The solution to any of the forms of the Euler equations provides the *answer*.

Of course this is only the briefest introduction to the variational calculus. We have discussed here only those elements that are directly relevant to problems that we will encounter subsequently in geometrical optics, such as rays in inhomogeneous media which follows next.

1.5 The Inhomogeneous Optical Medium

Now we apply the version of the Euler equations in Eq. 1.26 to the problem of rays in a medium in which the refractive index is a function of position [9] as indicated in Eqs. 1.4 and 1.5. Evidently $f(\mathbf{P},\, \mathbf{P}_t) = n(\mathbf{P})(ds/dt) = n(\mathbf{P})\sqrt{\mathbf{P}_t^2}$ which establishes f for this particular problem.

We must emphasize that in this context $\mathbf{P}(t)$ is a vector function representing all possible paths in the medium. Our problem is to find those particular paths that satisfy the Euler equations; those are the rays in this medium.

We cannot use s as the parameter in the statement of the variational problem because each possible arc will have a different geometrical length. A requirement for the application of these methods is that the parameter be uniform for all such curves. But s is not uniform so we must use a different parameter, say t, that *is* uniform over all possible arcs. This leads us to the following expressions

$$\nabla f = \sqrt{\mathbf{P}_t^2}\, \nabla n(\mathbf{P}), \qquad \nabla_t f = n(\mathbf{P}) \frac{\mathbf{P}_t}{\sqrt{\mathbf{P}_t^2}}. \tag{1.27}$$

Substituting these into Eq. 1.26, the vector form of the Euler equations, we get

$$\frac{d}{dt}\left(n(\mathbf{P})\, \frac{\mathbf{P}_t}{\sqrt{\mathbf{P}_t^2}} \right) = \sqrt{\mathbf{P}_t^2}\, \nabla n(\mathbf{P}). \tag{1.28}$$

But $ds/dt = \sqrt{\mathbf{P}_t^2}$ so that, reverting back to the arc length parameter s, Eq. 1.28 becomes

$$\frac{d}{ds}\left(n \frac{d\mathbf{P}}{ds} \right) = \nabla n. \tag{1.29}$$

This is the *ray equation* for an inhomogeneous medium. Provided that second derivatives exist, it can be expanded further

$$n\mathbf{P}_{ss} + (\nabla n \cdot \mathbf{P}_s)\mathbf{P}_s = \nabla n. \tag{1.30}$$

As always, we use subscripts to signal differentiation.

Equations 1.29 and 1.30 are ordinary differential equations for rays in a medium whose refractive index is a function of position and is continuous and differentiable in the variables $x,\ y,$ and z. An example of such is the *fish eye* of Maxwell which follows.

[9]Stavroudis 1972a, Chapter II. Luneburg 1964, pp. 164–172 discusses the special case where the medium has central symmetry.

1.6 The Maxwell Fish Eye

The ray equation, Eq. 1.29, works very well with Maxwell's fish eye.[10] The eye of a fish operates in water, a medium with a refractive index much higher than that of air, yet its lens is flat. This suggests that the eye of a fish, flat and immersed in a medium with a relatively high refractive index, has a low optical power implying a long back focal distance. Yet the flat structure includes the retina that then requires a short back focal distance. To explain away this paradox Maxwell postulated that the optical medium of the fish eye had a refractive index function in the following form

$$n(\mathbf{P}) = \frac{1}{1+\mathbf{P}^2}, \tag{1.31}$$

so that its gradient is

$$\nabla n = \frac{-2\mathbf{P}}{(1+\mathbf{P}^2)^2}. \tag{1.32}$$

Plugging this into the ray equation, Eq. 1.29, yields

$$\frac{d}{ds}\left[\frac{\mathbf{P}_s}{1+\mathbf{P}^2}\right] = \frac{-2\mathbf{P}}{(1+\mathbf{P}^2)^2}, \tag{1.33}$$

which quickly becomes

$$(1+\mathbf{P}^2)\mathbf{P}_{ss} - 2(\mathbf{P}\cdot\mathbf{P}_s)\mathbf{P}_s + 2\mathbf{P} = 0, \tag{1.34}$$

whose derivative is

$$(1+\mathbf{P}^2)\mathbf{P}_{sss} - 2(\mathbf{P}\cdot\mathbf{P}_{ss})\mathbf{P}_s = 0. \tag{1.35}$$

I do not know whether the fish eye is accurately described by this model or whether fish are even aware of the existence of these equations but as an example of an application of the Calculus of Variations to geometrical optics it will suffice.

We will contemplate these equations further in Chapter 2 which is concerned with the Differential Geometry of Space Curves.

1.7 The Homogeneous Medium

We can use Eq. 1.30 to handle the case where the refractive index n is a constant so that all its derivatives are zero. Then Eq. 1.30 degenerates to

$$\mathbf{P}_{ss} = 0, \tag{1.36}$$

a linear, ordinary differential of order *two* in vector form whose general solution must be

$$\mathbf{P}(s) = \mathbf{A}s + \mathbf{B}, \tag{1.37}$$

where $\mathbf{A}$ and $\mathbf{B}$ are vector constants of integration.

This is clearly a straight line showing us (as if we didn't already know!) that rays in homogeneous, isotropic media are, indeed, the shortest distance between two points.

[10] Luneburg 1964, pp. 172–182. Stavroudis 1972a, Chapter IV.

1.8 Anisotropic Media

In a certain sense the anisotropic medium is an analog of the inhomogeneous medium. In the latter medium the refractive index is a function of position and it can be represented by $n = n(\mathbf{P})$ while in the anisotropic medium it depends on a ray direction[11]. If $\mathbf{P}_s$ is a unit vector in the direction of a ray then it must be that $n = n(\mathbf{P}_s)$, superficially resembling the inhomogeneous medium but making an enormous difference in the variational integral and the Euler equations. Following Eq. 1.24 the variational integrand takes the form

$$f(\mathbf{P}_t) = n(\mathbf{P}_s)ds/dt = n\left(\frac{\mathbf{P}_t}{\sqrt{\mathbf{P}_t^2}}\right)\sqrt{\mathbf{P}_t^2}, \tag{1.38}$$

so that, in Eq. 1.26, $\nabla f = 0$ and the Euler equation becomes

$$\frac{d}{dt}\nabla_t f = \frac{d}{dt}\nabla_t\left[n\left(\frac{\mathbf{P}_t}{\sqrt{\mathbf{P}_t^2}}\right)\sqrt{\mathbf{P}_t^2}\right] = 0. \tag{1.39}$$

The leading component of the gradient is,

$$\begin{aligned}
\frac{\partial f}{\partial x_t} &= \frac{\partial}{\partial x_t}\left[n\left(\frac{\mathbf{P}_t}{\sqrt{\mathbf{P}_t^2}}\right)\sqrt{\mathbf{P}_t^2}\right] \\
&= \sqrt{\mathbf{P}_t^2}\left[\frac{\partial n}{\partial x_s}\frac{\mathbf{P}_t^2 - x_t^2}{(\mathbf{P}_t^2)^{3/2}} - \frac{\partial n}{\partial y_s}\frac{x_t\, y_t}{(\mathbf{P}_t^2)^{3/2}} - \frac{\partial n}{\partial z_s}\frac{x_t\, z_t}{(\mathbf{P}_t^2)^{3/2}}\right] + n\frac{x_t}{\sqrt{x_t^2}} \\
&= \frac{\partial n}{\partial x_s} - \frac{1}{\mathbf{P}^2}x_t\left[x_t\frac{\partial n}{\partial x_s} + y_t\frac{\partial n}{\partial y_s} + z_s\frac{\partial n}{\partial z_s}\right] + n\frac{x_t}{\sqrt{x_t^2}} \\
&= \frac{\partial n}{\partial x_s} - x_s\left[x_s\frac{\partial n}{\partial x_s} + y_s\frac{\partial n}{\partial y_s} + z_s\frac{\partial n}{\partial z_s}\right] + n x_s \\
&= \frac{\partial n}{\partial x_s} - x_s(\mathbf{P}_s \cdot \nabla_s n) + n x_s.
\end{aligned}$$

We do the same thing with the other two partial derivatives in $\nabla_t f$ to get

$$\left\{\begin{aligned}
\frac{\partial f}{\partial x_t} &= \frac{\partial n}{\partial x_s} - x_s(\mathbf{P}_s \cdot \nabla_s n) + n x_s \\
\frac{\partial f}{\partial y_t} &= \frac{\partial n}{\partial y_s} - y_s(\mathbf{P}_s \cdot \nabla_s n) + n y_s \\
\frac{\partial f}{\partial z_t} &= \frac{\partial n}{\partial z_s} - z_s(\mathbf{P}_s \cdot \nabla_s n) + n z_s,
\end{aligned}\right. \tag{1.40}$$

or, in vector form

$$\nabla_t f = \nabla_s n - \mathbf{P}_s(\mathbf{P}_s \cdot \nabla_s n) + \mathbf{P}_s n$$

[11] Avendaño-Alejo and Stavroudis 2002.

$$= \mathbf{P}_s \times (\nabla_s n \times \mathbf{P}_s) + \mathbf{P}_s n. \tag{1.41}$$

From Eq. 1.39 the derivative of Eq. 1.41 must vanish. It follows that there must exist a vector $\mathbf{A}$ that is independent of t (and therefore independent of s) so that

$$\mathbf{P}_s \times (\nabla_s n \times \mathbf{P}_s) + \mathbf{P}_s n = \mathbf{A}. \tag{1.42}$$

The scalar product of this with $\mathbf{P}_s$ yields

$$n = \mathbf{A} \cdot \mathbf{P}_s. \tag{1.43}$$

It follows from this that

$$\nabla_s n = \mathbf{A}. \tag{1.44}$$

This is about as far as we can go without making contact with physical reality; without taking into account the interaction of light with a physical medium.

From Eqs. 1.43 and 1.44 we can get

$$n = \nabla_s n \cdot \mathbf{P}_s, \tag{1.45}$$

a linear, first order partial differential equation that indicates that n must be a homogeneous function. But this is jumping the gun. We will show this and more in Chapter 5 on First Order Partial Differential Equations.

In this chapter we have covered a great deal of territory. We have studied the Calculus of Variations with fixed end points and its parametric representation and then on to a vector notation. This was then applied to inhomogeneous optical media inhomogeneous in general and to Maxwell's fish eye in particular. We have shown that in a homogeneous medium rays are straight lines. In a final brush with anisotropic media in we get inklings of some of the basic flaws in geometrical optics. But we also have laid some foundations on which will be erected new material in subsequent chapters.

2 Space Curves and Ray Paths

In the last chapter the Euler equations from the Calculus of Variations were applied to Fermat's principle. This led directly to the *ray equation* for an inhomogeneous medium, a system of second order *ordinary* differential equations.

Since these rays lie in an inhomogeneous medium they must be space curves and not straight lines and therefore in this chapter we will study curves in three dimensions. Once the general properties of these curves are developed they will then be applied to ray paths using the Euler equations, Eq. 1.29, that define them.

2.1 Space Curves

In Chapter 1, in the section on the parametric version of the Calculus of Variations, we introduced the idea of a vector function of a single parameter. We do the same thing here. Consider a vector as a function of a parameter t

$$\mathbf{P}(t) = \big(x(t),\ y(t),\ z(t)\big). \tag{2.1}$$

Note that the end point of $\mathbf{P}(t)$ sweeps out an arc as t varies. Figure 2.1 represents such a curve in three dimensions.

We will be concerned with the tangent to such a curve. The Euclidean definition of the tangent is that it is a straight line that touches a curve without crossing it. But consider a curve in the form of the letter S. In the upper part the tangent is clearly on the left side of the curve; in the lower part it is on the right. At some point the tangent sliding along the curve must change sides and at that point, the *inflection point*, it must cross the curve. At an inflection point the Euclidean definition of a tangent fails.

This paradox was resolved by Pierre de Fermat who defined the *tangent* by a limiting process. In his treatment, a chord connects two points, a and b, on the curve, as shown in Fig. 2.2. Hold a fixed and allow b to approach a along the curve. The limiting position of the chord is defined (by Fermat) to be the tangent to the curve at point a. It's quite possible that if Fermat had lived longer it would have been he who discovered the differential calculus. Consider the following.

Let the point a be represented by the vector $\mathbf{P}(t)$ and the point b by $\mathbf{P}(t + \Delta t)$. The difference, $\mathbf{P}(t + \Delta t) - \mathbf{P}(t)$, is the vector representing the chord connecting a and b. The limit of the difference quotient

$$\lim_{\Delta t \to 0} \frac{\mathbf{P}(t + \Delta t) - \mathbf{P}(t)}{\Delta t} = \frac{d\mathbf{P}}{dt}, \tag{2.2}$$

The Mathematics of Geometrical and Physical Optics: The k-function and its Ramifications. O.N. Stavroudis

ISBN: 3-527-40448-1

is exactly the derivative of $\mathbf{P}$ with respect to t, and at the same time, according to Fermat's definition, a vector tangent to the curve at point a.

We can go a little further using a more modern notation. Let $\Delta k = \sqrt{(\mathbf{P}(t+\Delta t) - \mathbf{P}(t))^2}\Delta t$ be the differential of the length of the chord. From the mean value theorem there exists a $\bar{t}$, so that $(\mathbf{P}(t + \Delta t) - \mathbf{P}(t) = \mathbf{P}_t(\bar{t})\Delta t$ where the subscript, as always, signals the derivative. It follows that

$$\Delta k = \sqrt{\mathbf{P}_t(\bar{t})^2}\,\Delta t. \tag{2.3}$$

Note that as b approaches a these two quantities approach each other so that we may write

$$\lim_{\Delta t \to 0} \frac{\Delta k}{\Delta s} = \lim_{\Delta t \to 0} \frac{\sqrt{(\mathbf{P}(t+\Delta t) - \mathbf{P}(t))^2}}{\Delta s} = 1. \tag{2.4}$$

Now we normalize the difference quotient in Eq. 2.4 making it a unit vector

$$\frac{\mathbf{P}(t+\Delta t) - \mathbf{P}(t)}{\sqrt{(\mathbf{P}(t+\Delta t) - \mathbf{P}(t))^2}} = \frac{\frac{\mathbf{P}(t+\Delta t) - \mathbf{P}(t)}{\Delta t}}{\frac{\sqrt{(\mathbf{P}(t+\Delta t) - \mathbf{P}(t))^2}}{\Delta t}} \tag{2.5}$$

whose limit as Δt approaches zero is

$$\frac{d\mathbf{P}}{dt} \Big/ \frac{ds}{dt} = \frac{d\mathbf{P}}{ds}, \tag{2.6}$$

the *unit* tangent vector. The parameter s is then the arc length that we had encountered in the last chapter; the derivative of $\mathbf{P}$ with respect to s is the *unit* tangent vector to the curve. By squaring both sides of Eq. 2.6 we find that

$$\frac{ds}{dt} = \sqrt{\left(\frac{d\mathbf{P}}{dt}\right)^2}. \tag{2.7}$$

Arc length is then a natural parameter to be used with space curves and in what follows it will be used exclusively; the derivative with respect to s will be denoted by the subscript $(_s)$. Recall that s is also the parameter most convenient for the *ray equation*, as indicated in Eqs. 1.29 and 1.30. These two equations will be used later on.

2.2 The Vector Trihedron

The derivative $\mathbf{P}_s$ is the unit tangent vector that we shall designate by

$$\mathbf{t} = \frac{d\mathbf{P}}{ds}. \tag{2.8}$$

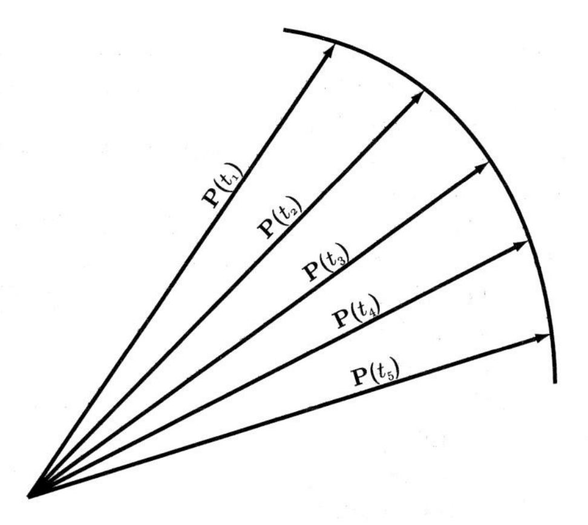

Figure 2.1: *The Vector Function Generates a Curve.*
$\mathbf{P}(t) = \big(x(t),\ y(t),\ z(t)\big)$ is a vector function. Each value of t produces a different vector $\mathbf{P}(t)$ originating at the coordinate origin. The locus of the end points is the generated curve.

From the fact that $\mathbf{t}^2 = 1$ it follows that $\mathbf{t} \cdot \mathbf{t}_s = 0$ so that $\mathbf{t}_s$ is perpendicular to $\mathbf{t}$. Define $\mathbf{n}$ as the *unit* normal vector in the direction of $\mathbf{t}_s$. Then we may write

$$\frac{d\mathbf{t}}{ds} = \mathbf{t}_s = \frac{1}{\rho}\,\mathbf{n}, \tag{2.9}$$

where $1/\rho$ is the rate of change of $\mathbf{t}$ as s increases. Now hold fixed the point $\mathbf{P}(s)$ (as well as s) and mark off the length ρ along the unit normal vector $\mathbf{n}$. For very small changes in s we can think of $\mathbf{t}$ as rotating around the point $\mathbf{P} + \rho\,\mathbf{n}$. The circle centered at this point with radius ρ is called the *osculating circle*; the plane determined by the vectors $\mathbf{t}$ and $\mathbf{n}$ is the *osculating plane*. At the point $\mathbf{P}(s)$ the curve and the osculating circle are tangent and also have a common radius of curvature ρ; therefore, $1/\rho$ is their *curvature*.

With $\mathbf{t}$ and $\mathbf{n}$ we define the unit *binormal* vector

$$\mathbf{b} = \mathbf{t} \times \mathbf{n}. \tag{2.10}$$

Analogous to the osculating plane, the *normal plane* is the plane determined by $\mathbf{n}$ and $\mathbf{b}$; the *rectifying plane*, by $\mathbf{t}$ and $\mathbf{b}$.

It is appropriate here to define a regular point on a curve as a point where $\mathbf{t}$, $\mathbf{n}$ and $\mathbf{b}$ are defined and are single valued, thus excluding points that are cusps or tack points or may otherwise cause embarrassment.

Now $\mathbf{t}$, $\mathbf{n}$ and $\mathbf{b}$ are unit vectors and are mutually perpendicular. These then constitute a set or orthonormal vectors, often referred to as the *vector trihedron*, [1] located at each point on the space curve, as shown in Fig. 2.3, and can be visualized as sliding along the curve as the value of the parameter s increases. We next find its properties.

[1] Struik 1961, pp. 19–23. The illustrations on pp. 19 and 21 are particularly useful.

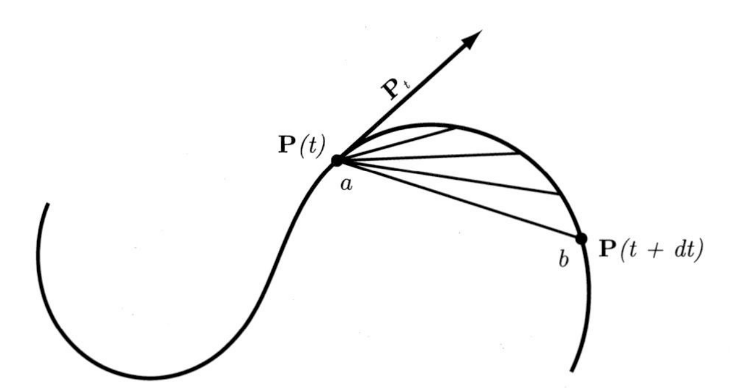

Figure 2.2: *The Derivative of a Vector Function is a Tangent.*
Let the endpoint of $\mathbf{P}(t)$ be point a; that of $\mathbf{P}(t+\Delta t)$ be point b. The difference between the two vectors is the chord $\alpha - b$. As $\Delta t \rightarrow 0$ point b approaches point a. The limiting position of the chord $a - b$ is the tangent at point a. The limit of the quotient formed by the vector difference and the arc length of the unit tangent vector to the curve at point a; this limit is also the derivative of the vector function $\mathbf{P}$ with respect to s.

2.3 The Frenet-Serret Equations

At every regular point on the curve the vector trihedron can serve as a local coordinate system; any vector through that point can be represented as a linear combination of $\mathbf{t}$, $\mathbf{n}$ and $\mathbf{b}$. We do this with the derivative of $\mathbf{n}$ so that

$$\mathbf{n_s} = \alpha\mathbf{t} + \beta\mathbf{n} + \gamma\mathbf{b}, \tag{2.11}$$

where the coefficients are to be determined. Since $\mathbf{n^2} = \mathbf{1}$, it follows that $\mathbf{n} \cdot \mathbf{n_s} = \mathbf{0}$; when we multiply Eq. 2.11 by $\mathbf{n}$ we see that $\beta = 0$. Since $\mathbf{t} \cdot \mathbf{n} = \mathbf{0}$, its derivative must also equal *zero*; $\mathbf{t} \cdot \mathbf{n_s} + \mathbf{t_s} \cdot \mathbf{n} = \mathbf{0}$. By multiplying Eq. 2.11 by $\mathbf{t}$ we get $\mathbf{t} \cdot \mathbf{n_s} = \alpha = -\mathbf{t_s} \cdot \mathbf{n} = -\mathbf{1}/\rho$. Since $\mathbf{n} \cdot \mathbf{b} = \mathbf{0}$ its derivative gives us $\mathbf{n_s} \cdot \mathbf{b} + \mathbf{n} \cdot \mathbf{b_s} = \mathbf{0}$ which results in $\mathbf{n_s} \cdot \mathbf{b} = \gamma = -\mathbf{n} \cdot \mathbf{b_s}$. With a similar calculation with $\mathbf{b} \cdot \mathbf{t}$ we get that $\mathbf{b_s} \cdot \mathbf{t} = -\mathbf{t_s} \cdot \mathbf{b} = \mathbf{1}/\rho$. From this and the nonvanishing of $\mathbf{n} \cdot \mathbf{b}_s$ we see that the change in $\mathbf{b}$ is also in the direction of $\mathbf{n}$; $\mathbf{b}$ therefore follows a twisting motion around the unit tangent vector $\mathbf{t}$ as it slides along the curve. We call this rate of twist the curve's *torsion* which we denote by $1/\tau$. Then, in Eq. 2.11, $\gamma = -1/\tau$. These results constitute the *Frenet-Serret equations*.[2] In matrix form they are

$$\begin{pmatrix} \mathbf{t}_s \\ \mathbf{n}_s \\ \mathbf{b}_s \end{pmatrix} = \begin{pmatrix} 0 & 1/\rho & 0 \\ -1/\rho & 0 & 1/\tau \\ 0 & -1/\tau & 0 \end{pmatrix} \begin{pmatrix} \mathbf{t} \\ \mathbf{n} \\ \mathbf{b} \end{pmatrix} \tag{2.12}$$

The vector trihedron slides along the space curve, much as an airplane glides along its flight path, moving in the direction of $\mathbf{t}$, a wing as its normal vector $\mathbf{n}$ and with $\mathbf{b}$ as its vertical stabilizer (or is it the other way around?), pitching and rolling as it moves along. The rate of roll is the torsion $1/\tau$; its pitch is the curvature $1/\rho$.

[2]Struik 1961. Chapter 1. Struik uses a different notation. κ is $1/\rho$ in our notation and τ is *torsion* rather than $1/\tau$.

2.4 When the Parameter is Arbitrary

Later on we will need to consider the case when the arbitrary parameter, which we will call t, cannot be transformed easily into the arc length parameter s. The relationship between these two parameters is determined by a total differential equation. If its solution is not obtainable or it is extremely complicated then it is necessary to refer $\mathbf{t}$, $\mathbf{n}$ and $\mathbf{b}$ back to the original parameter t.

Refer to Eqs. 2.6, 2.8 and 2.9 and get

$$\mathbf{t} = \mathbf{P}_s = \mathbf{P}_t/\sqrt{\mathbf{P}_t^2}, \tag{2.13}$$

from which comes

$$\begin{aligned}\mathbf{t}_s &= \frac{1}{\rho}\mathbf{n} = \frac{1}{\sqrt{\mathbf{P}_t^2}}\mathbf{t}_t = \frac{1}{(\mathbf{P}_t^2)^{3/2}}\left[(\mathbf{P}_t^2\mathbf{P}_{tt} - (\mathbf{P}_t \cdot \mathbf{P}_{tt})\mathbf{P}_t\right] \\ &= \frac{\mathbf{P}_t \times (\mathbf{P}_{tt} \times \mathbf{P}_t)}{(\mathbf{P}_t^2)^{3/2}},\end{aligned}$$

which reduces further to the curvature

$$\frac{1}{\rho} = \frac{1}{\mathbf{P}_t^2}\sqrt{(\mathbf{P}_{tt} \times \mathbf{P}_t)^2}, \tag{2.14}$$

and to the unit normal vector

$$\mathbf{n} = \frac{\mathbf{P}_t \times (\mathbf{P}_{tt} \times \mathbf{P}_t)}{\sqrt{\mathbf{P}_t^2}\sqrt{(\mathbf{P}_{tt} \times \mathbf{P}_t)^2}}. \tag{2.15}$$

We use Eqs. 2.17 and 2.19 to get the binormal vector

$$\mathbf{b} = \mathbf{t} \times \mathbf{n} = \frac{\mathbf{P}_t}{\sqrt{\mathbf{P}_t^2}} \times \frac{\mathbf{P}_t \times (\mathbf{P}_{tt} \times \mathbf{P}_t)}{\sqrt{\mathbf{P}_t^2}\sqrt{(\mathbf{P}_{tt} \times \mathbf{P}_t)^2}} = -\frac{\mathbf{P}_{tt} \times \mathbf{P}_t}{\sqrt{(\mathbf{P}_{tt} \times \mathbf{P}_t)^2}}. \tag{2.16}$$

Finally we come to the calculation of the torsion. For this we use the third equation of Eq. 2.12 for the derivative of $\mathbf{b}$ and take the derivative of Eq. 2.16 to get

$$\mathbf{b}_s = \frac{1}{\tau}\mathbf{n} = \frac{\mathbf{b}_t}{\sqrt{\mathbf{P}_t^2}} = -\frac{(\mathbf{P}_{ttt} \times \mathbf{P}_t) \cdot \mathbf{P}_{tt}}{\sqrt{\mathbf{P}_t^2}\left[(\mathbf{P}_{tt} \times \mathbf{P}_t)^2\right]^{3/2}}\mathbf{P}_t \times (\mathbf{P}_{tt} \times \mathbf{P}_t),$$

which, when compared with Eq. 2.15 yields

$$\frac{1}{\tau} = \frac{(\mathbf{P}_{ttt} \times \mathbf{P}_t) \cdot \mathbf{P}_{tt}}{(\mathbf{P}_{tt} \times \mathbf{P}_t)^2}. \tag{2.17}$$

These calculations give us curvature, torsion and the three orthogonal vectors of the trihedron in terms of the parameter t.

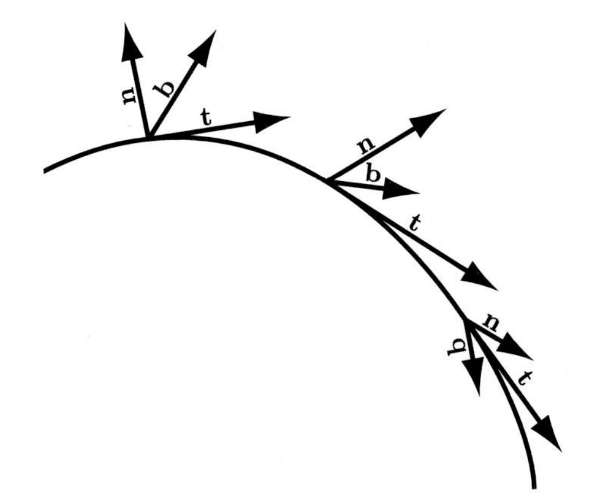

Figure 2.3: *The Vector Trihedron.* Showing the vectors $\mathbf{t}$, $\mathbf{n}$ and $\mathbf{b}$ at several points on the curve. Each forms a vector trihedron; this can be used as a local coordinate system.

2.5 The Directional Derivative

Recall the expression for the total differential of a function $f(\mathbf{V})$, as given in Chapter 1, from which we got the total derivative

$$\frac{df}{dt} = \nabla f \cdot \frac{d\mathbf{V}}{dt}. \tag{1.23}$$

We can interpret this as *the derivative of f in the direction of* $\mathbf{V}_t$, written as

$$f_t = (\mathbf{V}_t \cdot \nabla) f, \tag{2.18}$$

where we take $(\mathbf{V}_t \cdot \nabla)$ as the differential operator for the *directional derivative*.[3] This can be extended to vectors as follows:

$$\mathbf{W}_t = (\mathbf{V}_t \cdot \nabla)\mathbf{W}, \tag{2.19}$$

in which $(\mathbf{V}_t \cdot \nabla)$ is applied separately to each element of $\mathbf{W}$. This we read as *the derivative of* $\mathbf{W}$ *in the direction of* $\mathbf{V}_t$. Now the gradient of a vector is not easily defined; nevertheless the definition of the operator $(\mathbf{V}_t \cdot \nabla)$ presents no difficulties.

The Frenet-Serret equations, stated in terms of the directional derivatives, are

$$\begin{cases} (\mathbf{t} \cdot \nabla)\mathbf{t} = \dfrac{1}{\rho}\mathbf{n} \\[2ex] (\mathbf{t} \cdot \nabla)\mathbf{n} = -\dfrac{1}{\rho}\mathbf{t} + \dfrac{1}{\tau}\mathbf{b} \\[2ex] (\mathbf{t} \cdot \nabla)\mathbf{b} = -\dfrac{1}{\tau}\mathbf{n}. \end{cases} \tag{2.20}$$

These will be used in subsequent chapters.

[3] A more careful definition of the directional derivative or the *Gâteaux Variation* can be found in Troutman(1983), pp. 44 et seq.

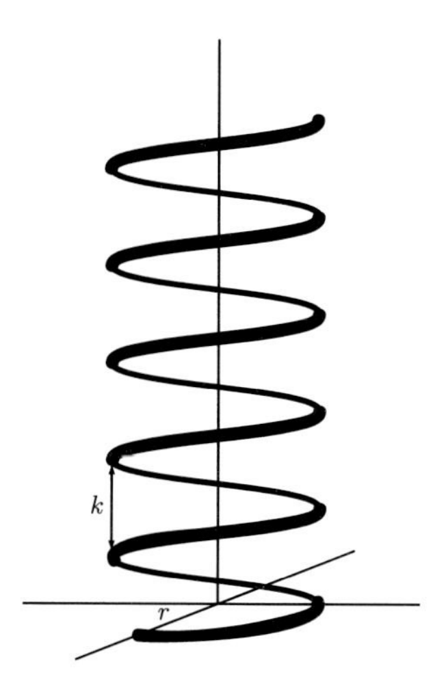

Figure 2.4: *The Cylindrical Helix.* $\mathbf{P}(t) = (r\cos t,\ r\sin t,\ kt)$ a spiral inscribed on a cylinder.

2.6 The Cylindrical Helix

To illustrate all this consider the helix, a curve inscribed on a circular cylinder of radius r. We will show that its curvature and torsion are constant. We take the axis of the cylinder to be along the z-axis so that a section parallel to the x, y-plane is a circle whose radius is r. The 'pitch' of the helix shall be the constant k so that in vector notation the appropriate equation is

$$\mathbf{P}(t) = (r\cos t,\ r\sin t,\ kt) \tag{2.21}$$

where t is the variable parameter.[4] This is shown in Fig. 2.4.

The arc length parameter s is therefore given by

$$(ds)^2 = \mathbf{P}_t^2 = (r^2 + k^2)(dt)^2, \tag{2.22}$$

which reduces to an exact total differential equation (to be treated in greater detail in Chapter 3) so that the relationship between s and t in finite terms is

$$t = \frac{s}{\sqrt{r^2 + k^2}}, \tag{2.23}$$

which lead us to the vector equation in terms of s

$$\mathbf{P}(s) = \left(r\cos\left(\frac{s}{\sqrt{r^2+k^2}}\right), r\sin\left(\frac{s}{\sqrt{r^2+k^2}}\right), \frac{ks}{\sqrt{r^2+k^2}}\right). \tag{2.24}$$

In what follows we will retain the use of the parameter t. The first derivative of **P** with respect to s is **t**

$$\mathbf{P}_s = \mathbf{t} = \frac{1}{\sqrt{r^2+k^2}}(-r\sin t,\ r\cos t,\ k). \tag{2.25}$$

[4]Stavroudis 1972a, pp. 33–35. A far more detailed study of the helix is in Struik 1961, pp. 33–35 with an illustration on p. 13.

Its derivative is

$$\mathbf{t}_s = \frac{1}{\rho}\mathbf{n} = \frac{-r}{\sqrt{r^2+k^2}}(\cos t,\ \sin t,\ 0), \tag{2.26}$$

so that the curvature must be

$$\frac{1}{\rho} = \frac{-r}{\sqrt{r^2+k^2}}, \tag{2.27}$$

and the unit normal vector

$$\mathbf{n} = (\cos t,\ \sin t,\ 0). \tag{2.28}$$

From Eqs. 2.25 and 2.28 we calculate the unit binormal vector

$$\mathbf{b} = \mathbf{t} \times \mathbf{n} = \frac{1}{\sqrt{r^2+k^2}}(-k\sin t,\ k\cos t, r) \tag{2.29}$$

the derivative of which leads to the torsion

$$\mathbf{b}_s = -\frac{1}{\tau}\mathbf{n} = \frac{-k}{\sqrt{r^2+k^2}}(\cos t,\ \sin t,\ 0), \tag{2.30}$$

so that

$$\frac{1}{\tau} = \frac{-k}{\sqrt{r^2+k^2}}. \tag{2.31}$$

Here we have used the Frenet-Serret equations, Eq. 2.17.

So, from Eqs. 2.27 and 2.31 we can see the curvature and torsion are indeed constants, exactly as advertised.

2.7 The Conic Section

We depart from the standard representation of the conic section in the following way. We write the equation of the ellipse in polar coordinates with a focus as the pole and the major axis as the polar axis. Then its equation will be given in parametric form by

$$\begin{cases} y = \varrho \sin\phi \\ z = \varrho\cos\phi \end{cases} \tag{2.32}$$

where the length of the radius vector is[5]

$$\varrho = \frac{r}{1 - \epsilon\cos\phi} \tag{2.33}$$

[5] Korn and Korn. pp. 41–53. Expressions for the conic section.

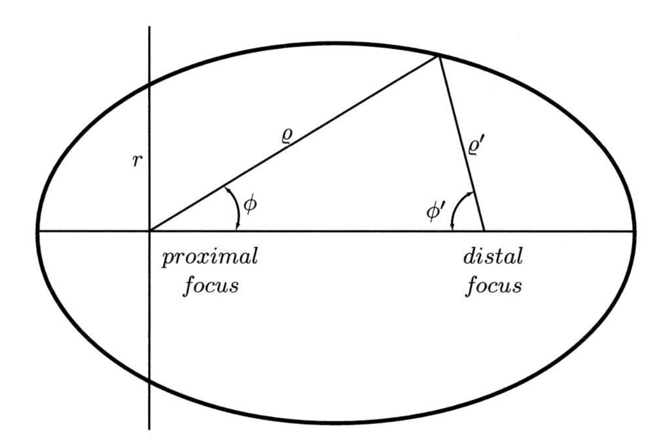

Figure 2.5: *The Ellipse and its Parameters.*
The ellipse is given in terms of polar coordinates with the pole located at a focus and with the polar axis lying along the major axis. We call the focus lying at the pole the *proximal focus*; the other, the *distal focus*. The eccentricity is $\epsilon < 1$. The angle υ is the parameter on which depends the radius vector, $\mathbf{R}(\upsilon) = r/(1 - \epsilon \cos \upsilon)$.

The *eccentricity* is $\epsilon \in (0,\ 1)$, the *semi latus rectum* is r. Because of the simplicity of Eq. 2.33 the use of eccentricity is preferable to the conic constant, the quantity usually encountered in optics.When $\upsilon = 0,\ \varrho$ is the distance from the coordinate origin, along the polar axis, to a vertex of the ellipse; $\varrho_0 = r/(1-\epsilon)$. When $\phi = \pi/2$, then ϱ equals the semi latus rectum; $\varrho_{\pi/2} = r$. Finally if $\phi = \pi$ then ϱ equals the distance along the polar axis in a negative direction to the ellipse's opposite vertex, $\varrho_\pi = r/(1+\epsilon)$.

We define the *proximal focus* of the ellipse as the first focus encountered; in this case it lies at the origin. The *distal focus* is defined as the second focus. These terms will be used extensively in Chapters 4 and 13. It follows from the results of the previous paragraph that the distance between the proximal and the distal foci is equal to the difference

$$d_f = \varrho_0 - \varrho_\pi = \frac{2\epsilon r}{1-\epsilon^2} \tag{2.34}$$

The length of the major axis is

$$2\,a = \varrho_0 + \varrho_\pi = \frac{2r}{1-\epsilon^2} \tag{2.35}$$

From the standard form of the ellipse we can derive the length of the semi minor axis as $b = r/\sqrt{1-\epsilon^2}$.

In Fig. 2.5 is shown an ellipse with these various quantities indicated. It also shows the angle ϕ' and the distance ϱ', analogues of ϕ and ϱ, but associated with the distal focus. From Eq. 2.33 it follows that

$$\varrho' = \frac{r}{1-\epsilon \cos \upsilon'} \tag{2.36}$$

so that, from Eq. 2.36, we can get

$$\begin{cases} y = \dfrac{r \sin v}{1 - \epsilon \cos v} = \dfrac{r \sin v'}{1 - \epsilon \cos v'} \\ z = 2\epsilon - (1 + \epsilon^2) \cos v = d_f - \dfrac{r \cos v'}{1 - \epsilon \cos v'}. \end{cases} \tag{2.37}$$

By substituting for d_f from Eq. 2.34 we are able to solve for v' to get

$$\begin{cases} \sin v' = (1 - \epsilon^2) \sin v / \mathcal{K}^2 \\ \cos v' = \left[2\epsilon - (1 + \epsilon^2) \cos v\right] / \mathcal{K}^2 \end{cases} \tag{2.38}$$

where

$$\mathcal{K}^2 = (1 + \epsilon^2) - 2\epsilon \cos v. \tag{2.39}$$

Note that $\mathcal{K}^2$ can be written as the sum of two squares, $(\cos v - \epsilon)^2 + \sin^2 v$, so that the quantity is never negative and $\mathcal{K}$ is always real.

From Eq. 2.37 we obtain easily

$$\varrho' = \frac{r\mathcal{K}^2}{(1 - \epsilon^2)(1 - \epsilon \cos v)}. \tag{2.40}$$

These same formulas, Eqs. 2.37 and 2.38, describe the circle ($\epsilon = 0$), the parabola ($\epsilon = 1$), and the hyperbola ($\epsilon > 1$), and therefore can be used to represent any conic section.

Now cast Eq. 2.37 into vector form

$$\mathbf{P}(v) = \varrho(0,\ \sin v,\ \cos v), \tag{2.41}$$

the derivative of which is

$$\mathbf{P}_v = \frac{r}{(1 - \epsilon \cos v)^2}(0,\ \cos v - \epsilon,\ -\sin v). \tag{2.42}$$

From Eq. 2.7 we can see that $\mathbf{P}_v^2 = (ds/dv)^2 = r^2\mathcal{K}^2/(1 - \epsilon \cos v)^4$, so that

$$\frac{ds}{dv} = \sqrt{\mathbf{P}_v^2} = \frac{r\mathcal{K}}{(1 - \epsilon \cos v)^2}. \tag{2.43}$$

Considered as an ordinary differential equation its solution involves an elliptic integral of the second kind, a relation too far complicated for our purposes. So we must be content with the results of the preceding section.Recall that **t**=$d\mathbf{P}/ds = (d\mathbf{P}/d\phi)/(ds/d\phi)$ so that

$$\mathbf{t} = \frac{1}{\mathcal{K}}(0,\ \cos v - \epsilon,\ -\sin v). \tag{2.44}$$

From $\mathbf{t}$ we can deduce the unit normal vector $\mathbf{n}$ and the curvature $1/\rho$ using Eqs. 2.20 and 2.44. First we calculate

$$\mathbf{t}_v = -\frac{1 - \epsilon \cos v}{\mathcal{K}^3}(0,\ \sin v,\ \cos v - \epsilon). \tag{2.45}$$

From this and Eq. 2.43 we get

$$\mathbf{t}_s = \frac{1}{\rho}\mathbf{n} = \frac{\partial \mathbf{t}/\partial v}{\partial s/\partial v} = -\frac{1-\epsilon\cos v}{\mathcal{K}^3}(0,\ \sin v,\ \cos v - \epsilon), \tag{2.46}$$

which lead us to

$$\frac{1}{\rho} = \frac{1}{r}\frac{(1-\epsilon\cos v)^2}{\mathcal{K}^3} \tag{2.47}$$

and

$$\mathbf{n} = \frac{1}{\mathcal{K}}(0,\ \sin v,\ \cos v - \epsilon). \tag{2.48}$$

From Eq. 2.10 we get

$$\mathbf{b} = \mathbf{t}\times\mathbf{n} = (1,\ 0,\ 0). \tag{2.49}$$

The explanation for this is that the ellipse is a plane curve that we have located on the y, z-plane. Its $\mathbf{b}$ vector must therefore be a unit vector in the x direction. It also follows that the curve's torsion $1/\tau$ must equal $zero$.

A caution; ρ and ϱ must not be confused. The radius of curvature of the ellipse is ρ and its curvature is $1/\rho$. The distance between a point on the ellipse and the proximal focus is ϱ; to its distal focus it is ϱ'. In Chapters 4 and 6 we will see that $1/\varrho$ and $1/\varrho'$ can also be interpreted as curvatures of spherical wavefronts which we will encounter in Chapter 13. These results will also be applied in Chapter 4, the chapter on surfaces, to get the equation for the prolate spheroid.

2.8 The Ray Equation

In the last chapter we derived the ray equation

$$\frac{d}{ds}\left(n\frac{d\mathbf{P}}{ds}\right) = \nabla n \tag{1.29}$$

or, expressed another way

$$n\mathbf{P}_{ss} + (\nabla n\cdot\mathbf{P}_s)\mathbf{P}_s = \nabla n. \tag{1.30}$$

Recall that all derivatives are with respect to the arc length parameter s. Expressing the ray equation as in Eq. 1.30, in terms of the tangent, normal and binormal vectors, we get

$$\frac{n}{\rho}\mathbf{n} + (\nabla n\cdot\mathbf{t})\mathbf{t} = \nabla n. \tag{2.50}$$

If we multiply this by $\mathbf{b}$ we see that

$$\nabla n\cdot\mathbf{b} = 0, \tag{2.51}$$

which says that the binormal vector of a ray path at any point in an inhomogeneous medium must always be perpendicular to that medium's gradient at that point. Another way of saying this is that the vectors $\mathbf{t}$, $\mathbf{n}$ and ∇n are coplanar.

We can rewrite Eq. 2.50 in the form

$$\frac{1}{\rho}\mathbf{n} = \frac{1}{n}\mathbf{t} \times (\nabla n \times \mathbf{t}) \tag{2.52}$$

by using the vector identity in Eq. A.6. Squaring this leads to an expression for the curvature of the ray at any point

$$\frac{1}{\rho} = \frac{1}{n}\sqrt{(\nabla n \times \mathbf{t})^2} \tag{2.53}$$

Taking the vector product of $\mathbf{t}$ with Eq. 2.52 we can get

$$\frac{n}{\rho}\mathbf{b} = -(\nabla n \times \mathbf{t}), \tag{2.54}$$

which, when differentiated with respect to s, leads to

$$\frac{n}{\rho\tau}\mathbf{n} + \left(\frac{n}{\rho}\right)_s \mathbf{b} = -\big[(\nabla n)_s \times \mathbf{t}\big] - \frac{1}{\rho}(\nabla n \times \mathbf{b}). \tag{2.55}$$

The scalar product of this with $\mathbf{n}$ yields to the expression for torsion at any point

$$\frac{1}{\tau} = \frac{1}{n}\big[\rho(\nabla n)_s \cdot \mathbf{b} + (\nabla n \cdot \mathbf{t})\big]. \tag{2.56}$$

2.9 More on the Fish Eye

In the last chapter we applied the Euler equations to Maxwell's fish eye and obtained the differential equation for rays in that medium

$$(1 + \mathbf{P}^2)\mathbf{P}_{ss} - 2(\mathbf{P} \cdot \mathbf{P}_s)\mathbf{P}_s + 2\mathbf{P} = 0, \tag{1.34}$$

which we then differentiate to get

$$(1 + \mathbf{P}^2)\mathbf{P}_{sss} - 2(\mathbf{P} \cdot \mathbf{P}_{ss})\mathbf{P}_s = 0. \tag{1.35}$$

The scalar product of this with $\mathbf{P}_{ss}$ yields $\mathbf{P}_{ss} \cdot \mathbf{P}_{sss} = 0$ which is the same as $d\mathbf{P}_{ss}^2/ds = 0$, from which $\mathbf{P}_{ss}^2 = \text{constant}$. From Eq. 2.9 we have that $\mathbf{P_{ss}} = \mathbf{t_s} = (\mathbf{1}/\rho)\mathbf{n}$ so that the curvature $(1/\rho)$ is constant along a ray

$$1/\rho = \mathit{constant}. \tag{2.57}$$

Moreover, by taking the scalar product of Eq. 1.35 with $\mathbf{P_s} \times \mathbf{P_{ss}}$, we get

$$\mathbf{P}_{sss} \cdot (\mathbf{P}_s \times \mathbf{P}_{ss}) = 0. \tag{2.58}$$

When this is compared with Eq. 2.17 we can see that the torsion is $zero$. Since the torsion is $zero$, rays in Maxwell's fish eye must be planar and since they have constant curvature they must be arcs of circles.

We can go a little further. Multiply Eq. 1.34 by $\mathbf{P}_{ss}$ (recall that $\mathbf{P}_s \cdot \mathbf{P}_{ss} = 0$) to get

$$2(\mathbf{P} \cdot \mathbf{P}_{ss}) = -(1 + \mathbf{P}^2)/\rho \tag{2.59}$$

When this is substituted into Eq. 1.35 we get

$$\mathbf{P}_{sss} + \frac{1}{\rho}\mathbf{P}_s = 0, \tag{2.60}$$

a linear, third order ordinary differential equation with constant coefficients. Its general solution is

$$\mathbf{P}(s) = \rho\big(\mathbf{A}\sin(s/\rho) - \mathbf{B}\cos(s/\rho) + \mathbf{C}\big), \tag{2.61}$$

where $\mathbf{A}$, $\mathbf{B}$, and $\mathbf{C}$ are vector constants of integration. Its derivative is

$$\mathbf{P}_s = \mathbf{A}\cos(s/\rho) + \mathbf{B}\sin(s/\rho) \tag{2.62}$$

The square of this must equal $unity$ yielding

$$\mathbf{A}^2\cos^2(s/\rho) + \mathbf{B}^2\sin^2(s/\rho) + 2(\mathbf{A} \cdot \mathbf{B})\sin(s/\rho)\cos(s/\rho) = 1, \tag{2.63}$$

from which we must conclude that $\mathbf{A}$ and $\mathbf{B}$ are both unit vectors and that $\mathbf{A} \cdot \mathbf{B} = 0$. The vectors $\mathbf{A}$ and $\mathbf{B}$ determine a plane. On this plane Eq. 2.61 provides the required solution, the arc of a circle.

What we have done in this chapter is to construct an analytic system for the study of space curves that, as we will see in a later chapter, is preliminary to the development of the differential geometry of surfaces. It is certainly incomplete. But we were able to apply it to rays paths in inhomogeneous media that led to a description of these objects in most general terms without needing to look at special cases or at boundary conditions for the resulting differential equation. We will return to this later when we apply this material to the pseudo Maxwell equations.

3 The Hilbert Integral and the Hamilton-Jacobi Theory

A few of the elementary problems in the Calculus of Variations were treated in Chapter 1 where it was assumed that the end points of the variational integrals were fixed. When these end points are allowed to vary the problem becomes much more difficult than merely solving the Euler equations. There had been no general method for solving this kind of problem but in the later years of the nineteenth century David Hilbert did find an approach that enabled us to treat this problem in a general way. This we call the *Hilbert Integral.*[1]

The Hilbert Integral has applications that transcend the variable end point problem; these will be our main interest in this chapter. With it we will find an alternative proof of Snell's law of refraction. We will find conditions for aggregates of rays to have orthogonal surfaces or wavefronts; such aggregates we will call *normal congruences* or *orthotomic systems.* The *Hilbert Integral* will also lead to the *Hamilton-Jacobi theory* that, in an optical context, is the foundation for modern aberration theory. It will also lead to the *eikonal equation* which will be of major importance in subsequent chapters.

Recall the statement of the general variational problem that had been introduced in Chapter 1

$$I = \int_a^b f\big(x,\, y(x),\, z(x),\, y_x(x),\, z_x(x)\big)dx, \tag{1.6}$$

and its solution in terms on the Euler equations

$$\begin{cases} \dfrac{d}{dx}\dfrac{\partial f}{\partial y_x} = \dfrac{\partial f}{\partial y} \\ \dfrac{d}{dx}\dfrac{\partial f}{\partial z_x} = \dfrac{\partial f}{\partial z}. \end{cases} \tag{1.15}$$

Also recall the steps we used, from Eq. 1.15, in Chapter 1, to get

$$\frac{\partial f}{\partial x} = \frac{d}{dx}\left[f - y_x\frac{\partial f}{\partial y_x} - z_x\frac{\partial f}{\partial z_x}\right]. \tag{1.17}$$

These are second order *ordinary* differential equations whose general solutions consist of a two parameter family of extremal curves that we will refer to as a *field of extremals.* In what follows we will assume that we are dealing with a region of space in which these extremals are dense.

[1]Bliss 1946, pp. 18–28.

The Mathematics of Geometrical and Physical Optics: The k-function and its Ramifications. O.N. Stavroudis

ISBN: 3-527-40448-1

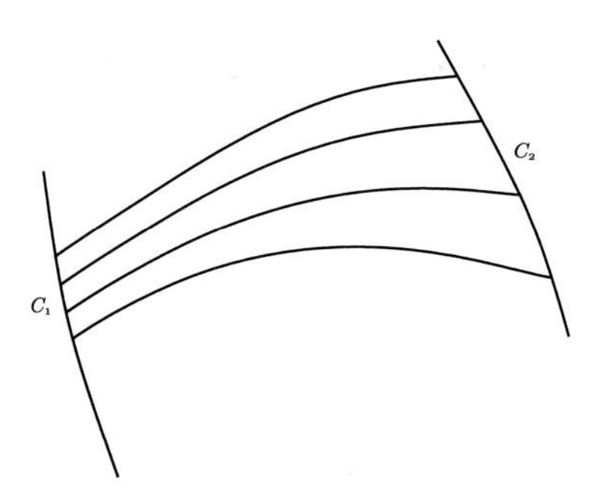

Figure 3.1: *The Hilbert Integral.* C_1 and C_2 represent two arbitrary surfaces connected by several rays.

Take two curves, not necessarily extremals, lying in such a field. These we represent by two sets of functions of a single parameter

$$C_1: \quad x_1(u),\ y_1(u),\ z_1(u); \qquad C_2: \quad x_2(u),\ y_2(u),\ z_2(u),$$

so each value of u determines two points, one on each curve. We can then form a variational integral connecting these two points, one on each curve, so that both are functions of u. See Fig. 3.1. Then the value of the variational integral is also a function of u

$$I(u) = \int_{x_1(u)}^{x_2(u)} f(x,\ y(x,u),\ z(x,u),\ y_x(x,u),\ z_x(x,u))dx. \tag{3.1}$$

Here the extremals, denoted by $y(x,u)$, $z(x,u)$, have their end points on C_1 and C_2 so that

$$\begin{aligned} y\big(x_1(u),u\big) &= y_1(u), z\big(x_1(u),u\big) = z_1(u),\\ y\big(x_2(u),u\big) &= y_2(u), z\big(x_2(u),u\big) = z_2(u). \end{aligned} \tag{3.2}$$

We next calculate the derivative of $I(u)$

$$\begin{aligned} \frac{dI}{du} &= f\big(x,\ y(x,u),\ z(x,u),\ y_x(x,u),\ z_x(x,u)\big)\Big|_{x=x_2(u)} \frac{dx_2}{du} \\ &\quad - f\big(x,\ y(x,u),\ z(x,u),\ y_x(x,u),\ z_x(x,u)\big)\Big|_{x=x_1(u)} \frac{dx_1}{du} \\ &\quad + \int_{x_1(u)}^{x_2(u)} \left[\frac{\partial f}{\partial y} y_u + \frac{\partial f}{\partial z} z_u + \frac{\partial f}{\partial y_x} y_{xu} + \frac{\partial f}{\partial z_x} z_{xu}\right] dx. \end{aligned} \tag{3.3}$$

The first two of these terms are written as one, $(dx/du)\, f\big|_{x=x_1}^{x=x_2}$. The integral is a bit more difficult. Using the Euler equations, Eq. 1.15, and the steps used in obtaining Eq. 1.15, we got

$$\int\limits_{x_1(u)}^{x_2(u)} \left[\frac{\partial f}{\partial y} y_u + \frac{\partial f}{\partial z} z_u + \frac{\partial f}{\partial y_x} y_{xu} + \frac{\partial f}{\partial z_x} z_{xu}\right] dx$$

$$= \int\limits_{x_1(u)}^{x_2(u)} \left[y_u \frac{d}{dx}\frac{\partial f}{\partial y_x} + z_u \frac{d}{dx}\frac{\partial f}{\partial z_x} + y_{xu}\frac{\partial f}{\partial y_x} + z_{xu}\frac{\partial f}{\partial z_x}\right] dx \tag{3.4}$$

$$= \int\limits_{x_1(u)}^{x_2(u)} \frac{d}{dx}\left[y_u \frac{\partial f}{\partial y_x} + z_u \frac{\partial f}{\partial z_x}\right] dx = \left[y_u \frac{\partial f}{\partial y_x} + z_u \frac{\partial f}{\partial z_x}\right]_{x=x_1(u)}^{x=x_2(u)}.$$

Next note that the *total* derivatives of y and z with respect to u are

$$\frac{dy}{du} = y_x \frac{dx}{du} + y_u\,; \qquad \frac{dz}{du} = z_x \frac{\partial x}{\partial u} + z_u. \tag{3.5}$$

Substituting for y_u and z_u in Eq. 3.4 yields an expression for dI/du,

$$\frac{dI}{du} = \left[f - y_x \frac{\partial f}{\partial y_x} - z_x \frac{\partial f}{z_x}\right]\frac{dx}{du} + \left[\frac{\partial f}{\partial y_x}\frac{dy}{du} + \frac{\partial f}{\partial z_x}\frac{dz}{du}\right]_{x=x_1(u)}^{x=x_2(u)}. \tag{3.6}$$

Now we define the *Hilbert Integral* along a curve C as,

$$J^*(C) = \int_{x_1}^{x_2} \left\{\left[f - y_x \frac{\partial f}{\partial y_x} - z_x \frac{\partial f}{z_x}\right]\frac{d\overline{x}}{du} + \frac{\partial f}{\partial y_x}\frac{d\overline{y}}{du} + \frac{\partial f}{\partial z_x}\frac{d\overline{z}}{du}\right\} du, \tag{3.7}$$

where $(\overline{x},\ \overline{y},\ \overline{z})$ lies on the path of integration. Then it follows that

$$\frac{dI}{du} = \frac{d}{du} J^*(C_2) - \frac{d}{du} J^*(C_1), \tag{3.8}$$

where $\bar{x}(u)$, $\bar{y}(u)$ and $\bar{z}(u)$ are the coordinate functions of the curve C. By integrating this equation between the limits u_1 and u_2 we get

$$I(E_b) - I(E_a) = J^*(C_2) - J^*(C_1), \tag{3.9}$$

where E_a and E_b are two extremals that correspond to the two values u_1 and u_2. This is shown in Fig. 3.2.

Another definition: If the integrand of the Hilbert Integral should vanish at a point then the curve C is said to be *transversal to the extremal at that point*; or more simply, *transversal at that point*. If the Hilbert integrand should vanish everywhere on C then C is said to be *transversal to the field of extremals*.

Note also that if the Hilbert Integral is expressed as a line integral its integrand takes the form

$$\left[f - y_x \frac{\partial f}{\partial y_x} - z_x \frac{\partial f}{\partial z_x}\right] d\overline{x} + \frac{\partial f}{\partial y_x} d\overline{y} + \frac{\partial f}{\partial z_x} d\overline{z}, \tag{3.10}$$

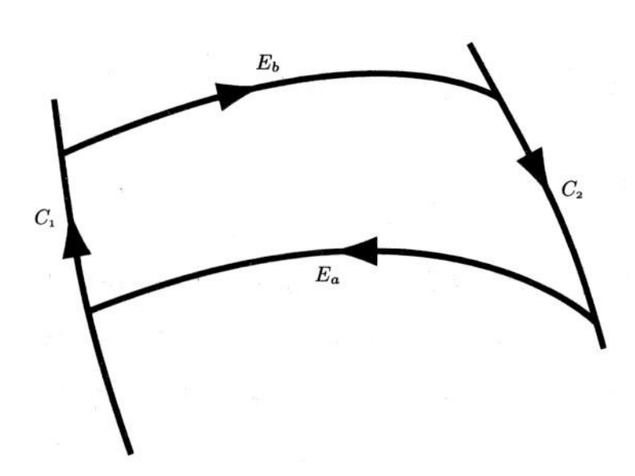

Figure 3.2: *The Hilbert Integral.* C_1 and C_2 are two surfaces connected by the pair of rays E_b and E_a. The Hilbert Integral, calculated around the closed path E_b, C_2, E_a, C_1, must equal *zero*.

a total differential. It follows that if a curve C is to be transversal to the field of extremals then a total differential equation must be satisfied. Going further, the existence of a solution to this total differential is a necessary and sufficient condition for the existence of a transversal.

3.1 A Digression on the Gradient

First of all a clarification. Some authors use ∇ as the differential operator whose components are partial derivatives with respect to a Cartesian coordinate system and use *grad*, *curl* and *div* to represent operators whose derivatives are with respect to different coordinate axes. Here we do not make that distinction; ∇ will denote such an operator under any coordinate system. For example, the gradient of a function $\mathcal{F}$, $\nabla\mathcal{F}$, will always represent a vector perpendicular to the surface $\mathcal{F} = constant$.

Now consider the integral of the gradient of $\mathcal{F}$, $\nabla\mathcal{F}$, along a curve $\mathbf{P}(t)(0 \le t \le 1)$ from the point $\mathbf{P}_0 = \mathbf{P}(0)$ to $\mathbf{P}_1 = \mathbf{P}(1)$

$$\int_{\mathbf{P}_0}^{\mathbf{P}_1} \nabla\mathcal{F} \cdot d\mathbf{P} = \int_0^1 \nabla\mathcal{F} \cdot \frac{d\mathbf{P}}{dt} dt.$$

This integrand can be written as a directional derivative as defined in Eq. 2.23 and the integral becomes

$$\int_0^1 (\mathbf{P}_t \cdot \nabla)\mathcal{F}\, dt = \int_0^1 \frac{d\mathcal{F}}{dt} dt = \mathcal{F}(\mathbf{P}_1) - \mathcal{F}(\mathbf{P}_0).$$

Notice that the value of the integral depends only on the value of $\mathcal{F}$ at the two end points and not on the path $\mathbf{P}(t)$ of the integral. We conclude that the integral of a gradient is independent of the path of integration. As a consequence, the integral of a gradient about a closed path must equal *zero*.[2]

$$\oint \nabla\mathcal{F} \cdot d\mathbf{P} = 0. \tag{3.11}$$

[2] Bliss 1946, Chapter III.

3.2 The Hilbert Integral. Parametric Case

Here we apply the vector notation in which the variational integral takes the form

$$I = \int_{t_1}^{t_2} f(\mathbf{P},\, \mathbf{P}_t)\, dt, \tag{1.24}$$

as shown in Chapter 1 so that the Euler equation becomes

$$\frac{d}{dt}\nabla_t f = \nabla f. \tag{1.26}$$

As before, let the two curves be $C_1 : \mathbf{V}(u)$ and $C_2 : \mathbf{W}(u)$ and construct a family of extremals, $\mathbf{P}(t,\ u)$, that connect the two. Then

$$\begin{aligned} &\mathbf{P}(t_1,\ u) = \mathbf{V}(u), \quad \mathbf{P}(t_2,\ u) = \mathbf{W}(u), \\ &\mathbf{P}_t(t_1,\ u) = \mathbf{V}_t(u), \mathbf{P}_t(t_2,\ u) = \mathbf{W}_t(u). \end{aligned} \tag{3.12}$$

Then the variational integral again becomes a function of u

$$I(u) = \int_{t_1}^{t_2} f\big(\mathbf{P}(t,\ u),\ \mathbf{P}_t(t,\ u)\big)dt, \tag{3.13}$$

and its derivative is then

$$\begin{aligned} \frac{dI}{du} &= \int_{t_1}^{t_2} \left[\nabla f \cdot \frac{d\mathbf{P}}{du} + \nabla_t f \cdot \frac{d\mathbf{P}_t}{du}\right] dt \\ &= \int_{t_1}^{t_2} \left[\left(\frac{d}{dt}(\nabla_t f)\right) \cdot \frac{\partial \mathbf{P}}{du} + \nabla_t f \cdot \left(\frac{d}{dt}\frac{\partial \mathbf{P}}{\partial u}\right)\right] \\ &= \int_{t_1}^{t_2} \frac{d}{dt}\left[\nabla_t f \cdot \frac{\partial \mathbf{P}}{\partial u}\right] dt = \left[\nabla_t f \cdot \frac{\partial \mathbf{P}}{\partial u}\right]_{t_1}^{t_2}. \end{aligned} \tag{3.14}$$

Expanding this out gives us

$$\begin{aligned} \frac{dI}{du} &= \nabla_t f\big(\mathbf{P}(t_2, u), \mathbf{P}_t(t_2, u)\big) \cdot \frac{\partial \mathbf{P}(t_2, u)}{\partial u} \\ &\quad - \nabla_t f\big(\mathbf{P}(t_1, u), \mathbf{P}_t(t_1, u)\big) \cdot \frac{\partial \mathbf{P}(t_1, u)}{\partial u} \\ &= \nabla_t f\big(\mathbf{W},\ \mathbf{W}_t\big) \cdot \frac{d\mathbf{W}}{du} - \nabla_t f\big(\mathbf{V},\ \mathbf{V}_t\big) \cdot \frac{d\mathbf{V}}{du}. \end{aligned} \tag{3.15}$$

Proceeding exactly as before, we integrate to get

$$I(u_2) - I(u_1) = J^*(\mathbf{W}) - J^*(\mathbf{V}), \tag{3.16}$$

where

$$J^*(\mathbf{V}) = \int_{u_1}^{u_2} \nabla_t f(\mathbf{V},\ \mathbf{V}_t) \cdot \mathbf{V}_u\, du, \tag{3.17}$$

is the Hilbert Integral over the curve $\mathbf{V}$. This also may be expressed as a line integral

$$J^*(\mathbf{V}) = \int_{\mathbf{V}_1}^{\mathbf{V}_2} \nabla_t f(\mathbf{V},\, \mathbf{V}_t) \cdot d\mathbf{V}. \tag{3.18}$$

The integrand in Eqs. 3.17 and 3.18 is then a gradient and therefore its integral around a closed curve must be zero as indicated in Eq. 3.11.

3.3 Application to Geometrical Optics

In the optical context the variational integral, given in Eq. 1.24, becomes

$$I = \int n(\mathbf{P})\sqrt{\mathbf{P}_t^2}dt, \tag{3.19}$$

so that

$$f = n(\mathbf{P})\sqrt{\mathbf{P}_t^2}, \qquad \nabla f = \sqrt{\mathbf{P}_t^2}\,\nabla n(\mathbf{P}), \qquad \nabla_t f = n(\mathbf{P})\frac{\mathbf{P}_t}{\sqrt{\mathbf{P}_t^2}}, \tag{3.20}$$

and the Euler equation becomes the ray equation, from Chapter 1,

$$\frac{d}{ds}\left(n\frac{d\mathbf{P}}{ds}\right) = \nabla n, \tag{1.29}$$

which expands to

$$n\mathbf{P}_{ss}+(\nabla n\cdot\mathbf{P}_s)\mathbf{P}_s = \nabla n. \tag{1.30}$$

The Hilbert Integral then becomes

$$J^*(\mathbf{V}) = \int n\mathbf{P}_s \cdot d\mathbf{V}. \tag{3.21}$$

The integrand vanishes when $\mathbf{P}_s \cdot d\mathbf{V} = 0$. Since $\mathbf{P}_s$ is tangent to the ray it must be that $\mathbf{V}$ is a surface perpendicular to the ray. In this case *transversality* becomes *orthogonality*. Setting the integrand equal to *zero* is then tantamount to forming a total differential equation for a surface that is perpendicular to the rays. The condition for the existence of a solution to the equation then becomes a condition on a family of rays for which such a surface exists. Such a system is called an *orthotomic system* or a *normal congruence*.

3.4 The Condition for Transversality

Let us return to the idea of transversality, the condition where the Hilbert Integral vanishes identically. As we can see in Eq. 3.21 in geometrical optics, a transversal curve intercepts a family of rays at right angles. In this case the Hilbert integrand becomes a total differential equation whose solution is a surface normal to the family of rays. In the next section we will obtain the conditions for such a differential to be *exact* or *differentiable* and how these conditions isolate families of rays for which an orthogonal surface exists. Then we will set these integrands equal to zero and consider the consequences.[3]

The three versions of the Hilbert integrand that we have seen are the most general case

(3.7) $$\left[f - y_x \frac{\partial f}{\partial y_x} - z_x \frac{\partial f}{z_x}\right] d\overline{x} + \frac{\partial f}{\partial y_x} d\overline{y} + \frac{\partial f}{\partial z_x} d\overline{z} = 0,$$

the parametric case

(3.18) $$\nabla_t f(\mathbf{V},\, \mathbf{V}_t) \cdot d\overline{\mathbf{V}} = 0,$$

and its application to geometric optics

(3.21) $$n\,(\mathbf{P}_s \cdot d\overline{\mathbf{V}}) = 0.$$

The barred variables pertain to coordinates on the transversal surface or, in the case of geometrical optics, on the orthogonal surface. The other component in each of the three equations refers to the field of extremals; the totality of solutions to the Euler equations. In what follows we will seek restrictions on the field of extremals that will assure the existence of transversals.

3.5 The Total Differential Equation

These are all a type of differential equation generally known as a *total differential equation*, a *linear differential form* or a *Pfaffian* and its most general form in three dimensions is[4]

$$P\,dx + Q\,dy + R\,dz = 0. \tag{3.22}$$

Lets assume that the solution of Eq. 3.22 is given in implicit form as, say $V(x,\, y,\, z) = 0$. Compare Eq. 3.22 with the total differential of V

$$dV = \frac{\partial V}{\partial x} dx + \frac{\partial V}{\partial y} dy + \frac{\partial V}{\partial z} dz = 0, \tag{3.23}$$

which gives us

$$\frac{\partial V}{\partial x} = P, \qquad \frac{\partial V}{\partial y} = Q, \qquad \frac{\partial V}{\partial z} = R. \tag{3.24}$$

If V is continuous then

$$\frac{\partial^2 V}{\partial x \partial y} = P_y = Q_x, \qquad \frac{\partial^2 V}{\partial x \partial z} = P_z = R_x, \qquad \frac{\partial^2 V}{\partial y \partial z} = Q_z = R_y, \tag{3.25}$$

[3] Stavroudis 1972a, pp. 73–76.
[4] Bliss 1946, pp. 70–72.

or

$$R_y - Q_z = 0, \qquad -R_x + P_z = 0, \qquad Q_x - P_y = 0. \tag{3.26}$$

If we write $\mathbf{H} = (P,\ Q,\ R)$ then it is easy to see that Eq. 3.26 is identical to

$$\nabla \times \mathbf{H} = 0, \tag{3.27}$$

which is a condition on $P,\ Q,$ and R that assures that Eq. 3.22 has a solution. This we call the condition for *exactness*; Eq. 3.22 is then said to be *exact*.

It can happen that Eq. 3.22 is not exact but that there exists an *integrating factor*, μ so that $\mu(Pdx + Qdy + Rdz) = 0$ is then exact. Then

$$\begin{aligned}
\frac{\partial^2 V}{\partial x \partial y} &= \mu P_y + \mu_y P = \mu Q_x + \mu_x Q, \\
\frac{\partial^2 V}{\partial x \partial z} &= \mu P_z + \mu_z P = \mu R_x + \mu_x R, \\
\frac{\partial^2 V}{\partial y \partial z} &= \mu Q_z + \mu_z Q = \mu R_y + \mu_y R,
\end{aligned} \tag{3.28}$$

which can be rearranged into

$$\begin{cases}
\mu(P_y - Q_x) = Q\mu_x - P\mu_y \\
\mu(P_z - R_x) = R\mu_x - P\mu_z \\
\mu(Q_z - R_y) = R\mu_y - Q\mu_z.
\end{cases} \tag{3.29}$$

By multiplying the first of these by R; the second, by $-Q$; and the third by P and adding we eliminate μ and its derivatives and get

$$P(R_y - Q_z) + Q(-R_x + P_z) + R(Q_x - P_y) = 0, \tag{3.30}$$

which we will call the *condition for integrability*. In vector form it is

$$\mathbf{H} \cdot (\nabla \times \mathbf{H}) = 0. \tag{3.31}$$

Now we apply this to the Hilbert integrand and refer back to Eq. 3.7

$$\mathbf{H} = \left(\left[f - y_x \frac{\partial f}{\partial y_x} - z_x \frac{\partial f}{z_x}\right],\ \frac{\partial f}{\partial y_x},\ \frac{\partial f}{\partial z_x}\right), \tag{3.32}$$

and to the application to geometrical optics, Eq. 3.21

$$\mathbf{H} = \nabla_t f(\mathbf{V},\ \mathbf{V}_t). \tag{3.33}$$

Later we will use Eq. 3.33 and the condition for integrability Eq. 3.31 to get the Hamilton-Jacobi equations on which modern dynamical mechanics and geometrical optics are based.

From the Euler equations we can show that if the equation is integrable then the index of refraction function, $n(\mathbf{P})$ is an integrating factor for Eq. 3.22. The condition for integrability is of course

$$\mathbf{P}_s \cdot (\nabla \times \mathbf{P}_s) = 0. \tag{3.34}$$

Note first of all that

$$\nabla \times (n\mathbf{P}_s) = \nabla n \times \mathbf{P}_s + n(\nabla \times \mathbf{P}_s). \tag{3.35}$$

By multiplying Eq. 3.34 by n we get the following:

$$\begin{aligned}\mathbf{P}_s \cdot [n(\nabla \times \mathbf{P}_s)] &= \mathbf{P}_s \cdot [\nabla \times (n\mathbf{P}_s) - \nabla \times \mathbf{P}_s] \\ &= \mathbf{t} \cdot \nabla \times (n\mathbf{P}_s) = 0.\end{aligned} \tag{3.36}$$

Take the derivative of this with respect to s to get

$$\begin{aligned}\frac{d}{ds}[\mathbf{t} \cdot \nabla \times (n\mathbf{P}_s)] &= \mathbf{t}_s \cdot \nabla \times (n\mathbf{P}_s) + \mathbf{t} \cdot \nabla \times \left[\frac{d}{ds}(n\mathbf{P}_s)\right] \\ &= \frac{1}{\rho}\mathbf{N} \cdot \nabla \times (n\mathbf{P}_s) + \mathbf{t} \cdot \nabla \times (\nabla n) = 0,\end{aligned} \tag{3.37}$$

where we have used Eq. 1.29. Since the curl of a gradient is zero this reduces to

$$\mathbf{N} \cdot \nabla \times (n\mathbf{P}_s) = 0. \tag{3.38}$$

Taking another derivative yields

$$\left[-\frac{1}{\rho}\mathbf{t} + \frac{1}{\tau}\mathbf{b}\right] \cdot [\nabla \times (n\mathbf{P}_s)] = 0. \tag{3.39}$$

whence comes

$$\mathbf{b} \cdot \nabla \times (n\mathbf{P}_s) = 0, \tag{3.40}$$

which shows that all three components of $\nabla \times (n\mathbf{P}_s)$ vanish so that the condition for exactness is satisfied.

These calculations form the basis for further relations which I have called the pseudo Maxwell equations which we will see in Chapter 13.

3.6 More on the Helix

Perhaps an example will make clear the use of the Hilbert Integral. Consider again the problem that we encountered in Chapter 2; finding the shortest distance from a fixed point to the cylindrical helix. To keep things simple we do not use the equation for the helix that involves

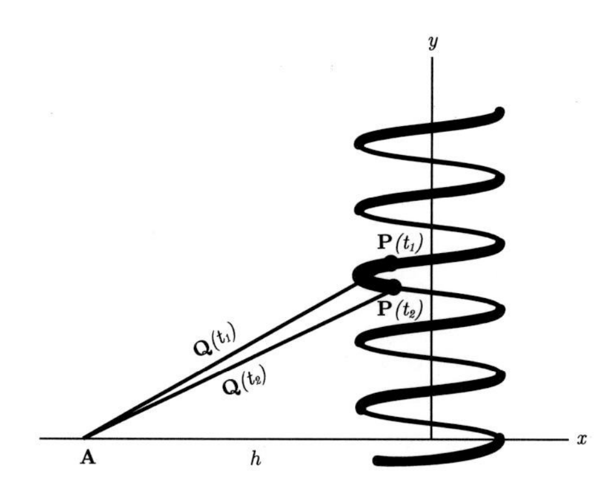

Figure 3.3: *The Helix and the Hilbert Integral.* Showing lines from a point $\mathbf{A}$ to the helix at $\mathbf{P}(t_1)$ and $\mathbf{P}(t_2)$. The Hilbert Integral about the closed loop $\mathbf{A}$, $\mathbf{P}(t_1)$, $\mathbf{P}(t_2)$, $\mathbf{A}$ is zero.

the arc length parameter s, Eq. 2.24, but the earlier version, Eq. 2.21, that uses only the arbitrary parameter t

$$\mathbf{P}(t) = (r\cos t,\, r\sin t,\, kt). \tag{2.21}$$

Place the fixed point $\mathbf{A}$ on the x-axis a distance h from the origin so that $\mathbf{A} = h\mathbf{X}$, as shown in Fig. 3.3. Then the vector from this point to the helix is

$$\mathbf{Q}(t) = h\mathbf{X} - \mathbf{P}(t), \tag{3.41}$$

whose length is

$$l(t) = \sqrt{\left(h\mathbf{X} - \mathbf{P}(t)\right)^2}. \tag{3.42}$$

The vector $\mathbf{Q}(t)$ is therefore a solution to the variational problem whose integral is $\int ds = \int (ds/dt)dt = \int \sqrt{\mathbf{Q}(t)^2}\, dt$. The Hilbert Integral that we will use will be based on this variational integral.

Now take two points on the helix corresponding to the two parameter values t_1 and t_2 and form the closed loop from $\mathbf{A}$ to $\mathbf{P}(t_1)$ to $\mathbf{P}(t_2)$ then back to $\mathbf{A}$. Because the loop is closed the Hilbert Integral calculated over this loop must be *zero*. Moreover, the Hilbert Integral calculated on $\mathbf{Q}(t_1)$ and $\mathbf{Q}(t_2)$, because they are extremals, degenerate to variational integrals, as shown in, say, Eq. 3.16. If we backtrack to Eq. 3.8, from which Eqs. 3.9 and 3.16 are derived, recalling that our loop has only three sides, we get

$$\frac{dI}{dt} = \frac{d}{dt} J^*\left(\mathbf{P}(t)\right). \tag{3.43}$$

Now note two things. First, for the distance from $\mathbf{A}$ to the helix to be a minimum the derivative of I must vanish. And second, the derivative of J^* is simply the integrand of the Hilbert

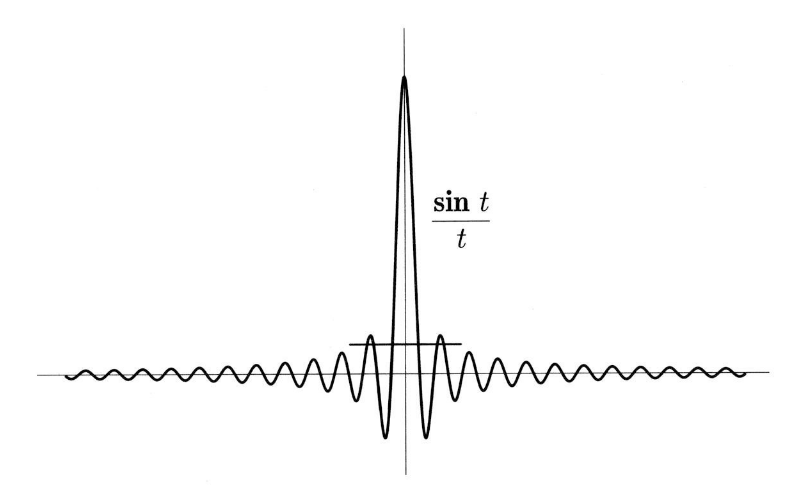

Figure 3.4: *A Graphical Solution.* The intersections of the parallel line with the curve are solutions to the equation $\frac{\sin t}{t} = -\frac{k^2}{hr}$.

Integral so that, from Eq. 3.18

$$\begin{aligned}
\nabla f \cdot d\mathbf{V} &= [\nabla \sqrt{(x-h)^2 + y^2 + z^2}] \cdot d\mathbf{P} \\
&= (1/f)(x-h,\ y,\ z) \cdot (x_t,\ y_t,\ z_t)dt \\
&= (1/f)\,(r\cos t - h,\ r\sin t,\ kt) \cdot (-r\sin t,\ r\cos t,\ k)\,dt \\
&= (1/f)(hr\sin t + k^2 t)dt = 0,
\end{aligned} \tag{3.44}$$

a transcendental equation in t. This can be solved graphically using

$$\frac{\sin t}{t} = -\frac{k^2}{hr}, \tag{3.45}$$

as shown in Fig. 3.4.

3.7 Snell's Law

We have already treated the variational integral for an inhomogeneous medium in Eq. 3.18 and from it we obtained the Hilbert Integral for such a medium in Eq. 3.21. Now suppose this medium contains a discontinuity; that there is a surface given by the equation $\mathcal{F}(\overline{\mathbf{P}}) = 0$ on which occurs a jump discontinuity of the refractive index function. By this we mean that on one side of the discontinuity the refractive index is continuous and is given by the function $n_1(\mathbf{P})$ and on the other side by $n_2(\mathbf{P})$, also continuous. As $\mathbf{P}$ approaches $\overline{\mathbf{P}}$ from the left the limit of $n_1(\mathbf{P})$ is $n_1(\overline{\mathbf{P}})$; as $\mathbf{P}$ approaches $\overline{\mathbf{P}}$ from the right the limit of $n_2(\mathbf{P})$ is $n_2(\overline{\mathbf{P}})$ and $n_1(\overline{\mathbf{P}}) \neq n_2(\overline{\mathbf{P}})$. This discontinuity, $\mathcal{F}(\overline{\mathbf{P}}) = 0$, is of course the refracting surface which we will call $\mathcal{S}$.

Take two points $\mathbf{A}$ and $\mathbf{B}$ on either side of the discontinuity and a point $\overline{\mathbf{P}}$ on the discontinuity and connect $\mathbf{A}$ and $\overline{\mathbf{P}}$ with a ray path $\mathbf{P}_1(s)$. Also connect $\overline{\mathbf{P}}$ and $\mathbf{B}$ with a second ray path $\mathbf{P}_2(s)$. These are, of course, extremals. Let I_1 and I_2 be the optical path lengths of the two curves determined by the appropriate variational integrals. Figure 3.5 is an attempt to show this. Let the 'prime' symbol indicate derivatives with respect to s. It follows that $\mathbf{P}_1'$ and $\mathbf{P}_2'$ are tangent vectors to the two curves and that $\overline{\mathbf{P}}_1'$ and $\overline{\mathbf{P}}_2'$ are their tangent vectors at the point of incidence on either side of the refracting surface. Then the total optical path length from $\mathbf{A}$ to $\mathbf{B}$ is given by

$$I = I_1 + I_2. \tag{3.46}$$

If $\mathbf{A}$ and $\mathbf{B}$ are not conjugates then the ray path connecting the two is unique and its optical path length must be an extremum relative to all other possible paths connecting the two points. Let $\mathbf{V}$ be any curve on $\mathcal{S}$. Then the differentials of I_1 and I_2 along $\mathbf{V}$ are

$$dI_1 = n_1\overline{\mathbf{P}}_1' \cdot d\mathbf{V}, \qquad dI_2 = -n_2\overline{\mathbf{P}}_2' \cdot d\mathbf{V}, \tag{3.47}$$

where $n_1 = n_1(\overline{\mathbf{P}})$ and $n_2 = n_2(\overline{\mathbf{P}})$. Each of these is a Hilbert integrand representing the differential of optical path length. The differential of the total optical path length between $\mathbf{A}$ and $\mathbf{B}$, a quantity that must equal zero in order to conform to Fermat's principle, is as follows:

$$dI = (n_1\overline{\mathbf{P}}_1' - n_2\overline{\mathbf{P}}_2') \cdot d\mathbf{V} = 0.$$

The differential $d\mathbf{V}$ is tangent to the curve $\mathbf{V}$. It follows that the vector $n_1\overline{\mathbf{P}}_1' - n_2\overline{\mathbf{P}}_2'$ must be perpendicular to $\mathbf{V}$. This must be true for all curves $\mathbf{V}$ on passing through $\overline{\mathbf{P}}$ on $\mathcal{S}$. It follows that $n_1\overline{\mathbf{P}}_1' - n_2\overline{\mathbf{P}}_2'$ must be parallel to the unit normal vector to $\mathcal{S}$ at $\overline{\mathbf{P}}$. Let this vector be $\mathbf{N}$. Then, for some factor k it must be that $n_1\overline{\mathbf{P}}_1 - n_2\overline{\mathbf{P}}_2 = k\mathbf{N}$ or

$$(n_1\overline{\mathbf{P}}_1' - n_2\overline{\mathbf{P}}_2') \times \mathbf{N} = 0,$$

which becomes very quickly the vector form of Snell's law

$$n_1(\overline{\mathbf{P}}_1' \times \mathbf{N}) = n_2(\overline{\mathbf{P}}_2' \times \mathbf{N}). \tag{3.48}$$

Recall that the absolute value of the vector product of two unit vectors is equal to the sine of the subtended angle so that $|\overline{\mathbf{P}}_1' \times \mathbf{N}| = \sin i$ and $|\overline{\mathbf{P}}_2' \times \mathbf{N}| = \sin r$, i and r being the angles of incidence and refraction. This leads to the scalar form of Snell's law

$$n_1 \sin i = n_2 \sin r, \tag{3.49}$$

to which must be added the fact that $\overline{\mathbf{P}}_1'$, $\overline{\mathbf{P}}_2'$ and $\mathbf{N}$ must be coplanar. This comes from Eq. 3.49. This plane is called the *plane of incidence*.

A matter of notation. In Chapter 7 in which ray tracing is studied we will use a different notation. Equation 3.48 then becomes

$$n_1(\mathbf{S}_1 \times \mathbf{N}) = n_2(\mathbf{S}_2 \times \mathbf{N}). \tag{3.50}$$

We will use these results in subsequent chapters but in a slightly different form. We will use the *reduced* direction cosine vector or the *reduced* ray vector $\mathbf{S} = n\,\mathbf{P}$ so that Eq. 3.48 becomes

$$\mathbf{S}_1 \times \mathbf{N} = \mathbf{S}_2 \times \mathbf{N}. \tag{3.51}$$

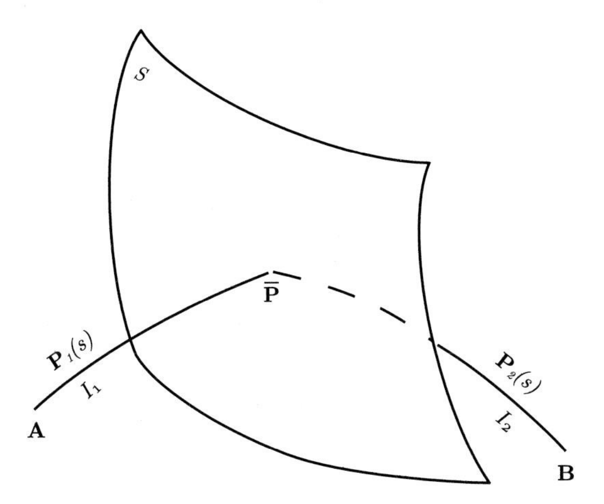

Figure 3.5: *Conjugate Points.* The ray path from **A** to **B** is unique unless **A** and **B** are perfect conjugates. I_1 and I_2 represent lengths from **A** to $\overline{\mathbf{P}}$ and from $\overline{\mathbf{P}}$ to **B**. Refraction occurs at $\overline{\mathbf{P}}$.

3.8 The Hamilton-Jacobi Partial Differential Equations

Recall the general variational integral Eq. 1.6, the associated Euler equations Eq. 1.15 from which comes

$$\frac{d}{dx}\left[f - y_x\frac{\partial f}{\partial y_x} - z_x\frac{\partial f}{\partial z_x}\right] = -\frac{\partial f}{\partial x}. \tag{1.17}$$

Also, we will need the Hilbert Integral Eq. 3.7 in which $\bar{x}, \bar{y}$ and $\bar{z}$ represent coordinates on a surface that intersects each extremal in the field. The other quantities in the Hilbert Integral are associated with the field of extremals. We are to find conditions on the field of extremals that assure the existence of transversals.

What follows is due to Hamilton.[5]

First, define the *canonical variables* u and v as follows,

$$u = \frac{\partial f}{\partial y_x}, \qquad v = \frac{\partial f}{\partial z_x}. \tag{3.52}$$

Recall that the function f is in the form $f(x,\ y,\ z,\ y_x,\ z_x)$. Consider the two definitions in Eq. 3.52 a pair of simultaneous equations in y_x and z_x which can be solved, to yield

$$y_x = P(x,\ y,\ z,\ u,\ v), \qquad z_x = Q(x,\ y,\ z,\ u,\ v). \tag{3.53}$$

Next, define the *Hamiltonian* (recall Eq. 1.17)

$$H(x,\ y,\ z,\ u,\ v) = y_x\frac{\partial f}{\partial y_x} + z_x\frac{\partial f}{\partial z_x} - f. \tag{3.54}$$

[5] Carathéodory 1999, pp. 121–133.

Using Eqs. 3.52 and 3.53 it is clear that this can also be written as

$$H(x,\, y,\, z,\, u,\, v) = Pu + Qv - f. \tag{3.55}$$

The total differential of the Hamiltonian is

$$dH = \frac{\partial H}{\partial x}dx + \frac{\partial H}{\partial y}dy + \frac{\partial H}{\partial z}dz + \frac{\partial H}{\partial u}du + \frac{\partial H}{\partial v}dv, \tag{3.56}$$

but by using Eq. 3.55 we get

$$\begin{aligned} dH &= P\,du + u\,dP + Q\,dv + v\,dQ \\ &\quad - \left[\frac{\partial f}{\partial x}dx + \frac{\partial f}{\partial y}dy + \frac{\partial f}{\partial z}dz + \frac{\partial f}{\partial y_x}dy_x + \frac{\partial f}{\partial z_x}dz_x\right] \\ &= P\,du + u\,dP + Q\,dv + v\,dQ \\ &\quad - \left[\frac{\partial f}{\partial x}dx + \frac{\partial f}{\partial y}dy + \frac{\partial f}{\partial z}dz + u\,dP + v\,dQ\right] \\ &= Pdu + Qdv - \left[\frac{\partial f}{\partial x}dx + \frac{\partial f}{\partial y}dy + \frac{\partial f}{\partial z}dz\right], \end{aligned} \tag{3.57}$$

where we have used Eq. 3.53. By equating coefficients of corresponding differentials we get

$$\frac{\partial H}{\partial x} = -\frac{\partial f}{\partial x}, \tag{3.58}$$

which we already have from Eq. 1.17, and

$$\begin{aligned} \frac{\partial H}{\partial y} &= -\frac{\partial f}{\partial y} = -\frac{d}{dx}\frac{\partial f}{\partial y_x} = \frac{du}{dx}, \qquad & \frac{\partial H}{\partial u} &= P = y_x = \frac{dy}{dx}, \\ \frac{\partial H}{\partial z} &= -\frac{\partial f}{\partial z} = -\frac{d}{dx}\frac{\partial f}{\partial z_x} = \frac{dv}{dx}, \qquad & \frac{\partial H}{\partial v} &= Q = z_x = \frac{dz}{dx}, \end{aligned} \tag{3.59}$$

where we have used Eqs. 3.52 and 3.53. These are called the *canonical equations*.

Now return to the Hilbert integrand from Eq. 3.7. When this is set equal to zero we have the total differential equation for the transversal surface

$$-Hd\bar{x} + ud\bar{y} + vd\bar{z} = 0. \tag{3.60}$$

The first two of the canonical equations, Eq. 3.59, satisfy the condition that Eq. 3.60 be exact. Then there must exist a solution to Eq. 3.60, a function ϕ such that

$$\frac{\partial \phi}{\partial \bar{x}} = -H(x,\, y,\, z,\, u,\, v), \qquad \frac{\partial \phi}{\partial \bar{y}} = u, \qquad \frac{\partial \phi}{\partial \bar{z}} = v. \tag{3.61}$$

By elimination u and v from the first equation in Eq. 3.61 yields the Hamilton-Jacobi partial differential equation for the transversal surface

$$\frac{\partial \phi}{\partial \bar{x}} + H\left(x,\, y,\, z,\, \frac{\partial \phi}{\partial \bar{y}},\, \frac{\partial \phi}{\partial \bar{z}}\right) = 0. \tag{3.62}$$

3.9 The Eikonal Equation

This is much simpler. Recall the Hilbert integrand expressed in the context of geometrical optics, Eq. 3.21

$$n\mathbf{P}_s \cdot d\mathbf{V} = 0,$$

and its condition for exactness from Eq. 3.27

$$\nabla \times (n\mathbf{P}_s) = 0. \tag{3.63}$$

Since the curl of a gradient always equals *zero* there must exist a function ϕ such that

$$n\mathbf{P}_s = \nabla\phi, \tag{3.64}$$

where $\phi(\mathbf{P}) = \text{constant}$ is the equation of the transversal surface or, in this case, the orthogonal surface, the plane of constant phase or wavefront. The function ϕ is referred to as *Hamilton's characteristic function* or, as I prefer, the *eikonal*. Recall that $\mathbf{P}_s$ is a unit vector in the direction of a ray; it is also the normal to the wavefront.

Now square both sides of Eq. 3.64 to get the *eikonal equation*, the analog to the Hamilton-Jacobi equation

$$(\nabla\phi)^2 = n^2. \tag{3.65}$$

Note that it is a non-linear partial differential equation of the first order and that it is *independent of any reference to rays.*

A solution to this equation, of course, would be very useful in optical design. Such a solution in the form of a power series expansion has been used in lens design for the better part of a century. Moreover it has the property that the zero order terms are exactly equivalent to the Gauss-Maxwell model for a perfect lens that we will encounter in Chapter 14. The coefficients of the terms of the power series therefore represent departures from the ideal and are termed *aberrations*. These terms are calculated and are shown to depend on the various parameters of the optical system being designed; curvatures of the refracting surfaces, separations between these surfaces and the refractive indices of the constituent media.

We will depart from the power series approach and find a general solution to Eq. 3.65 in finite terms in Chapter 5.

4 The Differential Geometry of Surfaces.

Any smooth surface (*Smooth* will be defined shortly) can be represented by a vector function in *two* parameters just as a space curve can be defined by a vector function of *one*. Indeed we will use space curves embedded on such a surface to reveal its properties. The ultimate objective is to apply these results to wavefront trains in homogeneous, isotropic media, whose equations will be found in Chapter 5 as a general solution of the eikonal equation derived in Chapter 3.

In Chapter 6 the results obtained in this chapter will be used to determine the geometric properties of these wavefronts as well as expressions for the associated caustic surfaces.

4.1 Parametric Curves[1]

A surface is determined completely by a vector function of two parameters, $\mathbf{P}(v,\ w)$. If we hold v fixed and allow w to vary the result is a space curve embedded on the surface. More accurately, $\mathbf{P}(v_0,\ w)$ is a one-parameter family of curves, with v_0 as the parameter, etched on this surface, just as $\mathbf{P}(v,\ w_0)$ is a second one-parameter family of curves lying on the surface. Together, these two families of curves are referred to as the *parametric curves* of the surface such as is shown in Fig. 4.1. Through every *regular* point (Again, we postpone the definition of a term.) passes exactly two parametric curves, one from each of these two families.

These parameters are not unique. While a nonsingular transformation of the parameters results in a different set of parametric curves on the surface the surface itself remains unchanged. We have already seen this in space curves; a change of parameterization does not change the geometry of the curve in any way. In that case we settled on a standard parameter s, the arc length parameter, as natural and convenient. However, parameters in the vector representation of surface are not standardized.

4.2 Surface Normals

At every regular point on the surface two of these parametric curves will intersect. The partial derivatives of $\mathbf{P}$ with respect to the two parameters v and w, as we saw in Chapter 2, are two vectors tangent to the two parametric curves and therefore tangent to the surface. These we denote by $\mathbf{P}_v$ and $\mathbf{P}_w$, where the subscripts signal partial differentiation.

Now we can define *smooth*. We say that a surface is smooth at a point if its partial derivatives of its representative function exist, are continuous and are single valued. In addition, on a

[1] My favorite text in classical differential geometry, the kind used in this work, is Struik 1961. Others include Stoker 1969, Guggenheimer 1977 and Do Carmo 1976. And perhaps the best of all is Blaschke 1945.

The Mathematics of Geometrical and Physical Optics: The k-function and its Ramifications. O.N. Stavroudis

ISBN: 3-527-40448-1

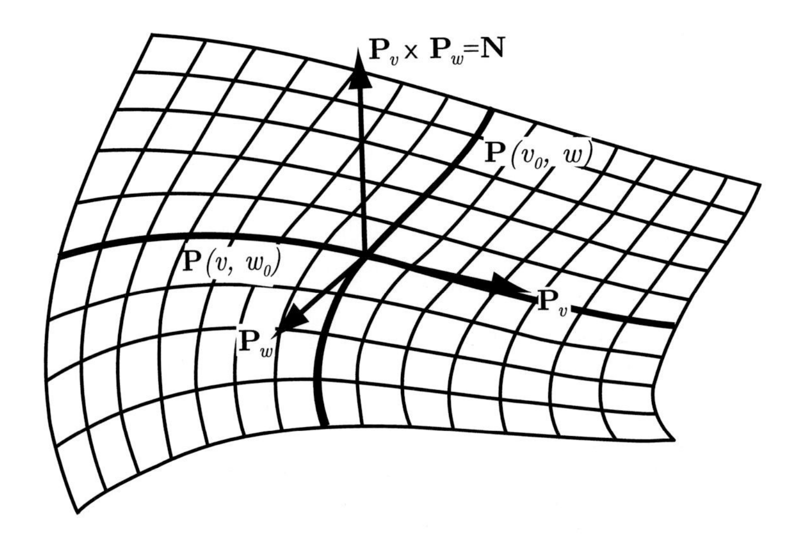

Figure 4.1: *Parametric Curves.* Shown are $\mathbf{P}(v,\ w_0)$ and $\mathbf{P}(v_0,\ w)$, the two one-parameter families of parametric curves. $\mathbf{P}_v$ and $\mathbf{P}_w$ are the tangent vector to these two systems of curves and $\mathbf{P}_v \times \mathbf{P}_w$ is the surface normal, $\mathbf{N}$.

smooth surface if the vector product $\mathbf{P}_v \times \mathbf{P}_w$ does not vanish at some point then that point is *regular*. If all points on a smooth surface are regular then the surface is said to be *regular*.

As an example, consider a sphere described by parametric curves that are lines of latitude and longitude. The longitude curves are great circles. Through almost all points of the sphere pass exactly two of the parametric curves; partial derivatives of the vector function describing the sphere exist and are unique so that there is no problem with the surface normal. However at the north and south poles all longitudinal curves intersect and the curve of latitude becomes a degenerate point circle. The two poles then can not be regular points. These two singularities can be removed by making a change of parameters but if lines of curvature are to be lines of latitude and longitude then all that is accomplished is the relocation of the two singularities to a different pair of points.

Since $\mathbf{P}_v$ and $\mathbf{P}_w$ are tangents to the surface it follows that at every regular point their vector product is a vector perpendicular to the surface. We then define the *unit normal vector* as

$$\mathbf{N} = \frac{\mathbf{P}_v \times \mathbf{P}_w}{\sqrt{(\mathbf{P}_v \times \mathbf{P}_w)^2}}. \tag{4.1}$$

This is shown in Fig. 4.1. Since $\mathbf{N}$ is a surface property it must be invariant under any parameter transformation.

Now let v and w be functions of a new parameter t; that is, let $v = v(t)$ and $w = w(t)$. A new space curve results, one that is engraved on the surface and is defined by the vector function $\mathbf{P}(t) = \mathbf{P}\big(v(t),\ w(t)\big)$ and can be analyzed using methods from the differential geometry of space curves that we encountered in Chapter 2. The tangent vector is

$$\mathbf{P}_t = \mathbf{P}_v v_t + \mathbf{P}_w w_t, \tag{4.2}$$

a linear combination of the two vectors tangent to the parametric curves.

The differential of arc length is

$$\begin{aligned}(ds)^2 &= [\mathbf{P}_v\, v_t + \mathbf{P}_w\, w_t]^2\, (dt)^2 \\ &= \mathbf{P}_v^2\,(dv)^2 + (\mathbf{P}_v \cdot \mathbf{P}_w)\, dv\, dw + \mathbf{P}_w^2\,(dw)^2.\end{aligned} \tag{4.3}$$

We define the *first fundamental quantities* to be

$$E = \mathbf{P}_v^2, \qquad F = \mathbf{P}_v \cdot \mathbf{P}_w, \qquad G = \mathbf{P}_w^2. \tag{4.4}$$

These are independent of t and so they are properties of the surface and not of the inscribed curve. Then the derivative of the arc length parameter s with respect to the parameter t becomes

$$\frac{ds}{dt} = \sqrt{E\, v_t^2 + 2\, F\, v_t\, w_t + G\, w_t^2}. \tag{4.5}$$

The radicand $E\, v_t^2 + 2\, F\, v_t\, w_t + G\, w_t^2$ is called the *first fundamental form*. This enables us to calculated derivatives with respect to s, the arc length parameter. The *unit* tangent vector then becomes

$$\mathbf{t} = \mathbf{P}_s = \frac{\mathbf{P}_v\, v_t + \mathbf{P}_w\, w_t}{\sqrt{E\, v_t^2 + 2\, F\, v_t\, w_t + G\, w_t^2}} = \mathbf{P}_v\, v_s + \mathbf{P}_w\, w_s. \tag{4.6}$$

The surface unit normal vector can now be written as

$$\mathbf{N} = (\mathbf{P}_v \times \mathbf{P}_w)/D, \tag{4.7}$$

where

$$D^2 = (\mathbf{P}_v \times \mathbf{P}_w)^2 = \mathbf{P}_v^2 \mathbf{P}_w^2 - (\mathbf{P}_v \cdot \mathbf{P}_w)^2 = EG - F^2. \tag{4.8}$$

4.3 The Theorem of Meusnier[2]

We next take the derivative of the unit tangent vector with respect to the arc length parameter and obtain the unit normal vector and the curvature, as in the first of the three Frenet-Serret equations in Chapter 2

$$\mathbf{t}_s = \frac{1}{\rho}\mathbf{n} \tag{2.20}$$

from which we get

$$\mathbf{t}_s = \frac{1}{\rho}\mathbf{n} = \mathbf{P}_{vv} v_s^2 + 2\, \mathbf{P}_{vw} v_s w_s + \mathbf{P}_{ww} w_s^2 + \mathbf{P}_v v_{ss} + \mathbf{P}_w w_{ss}. \tag{4.9}$$

[2]Struik 1961, pp. 73–76, Do Carma 1976, pp. 142–144, Korn and Korn 1968, pp. 571–572.

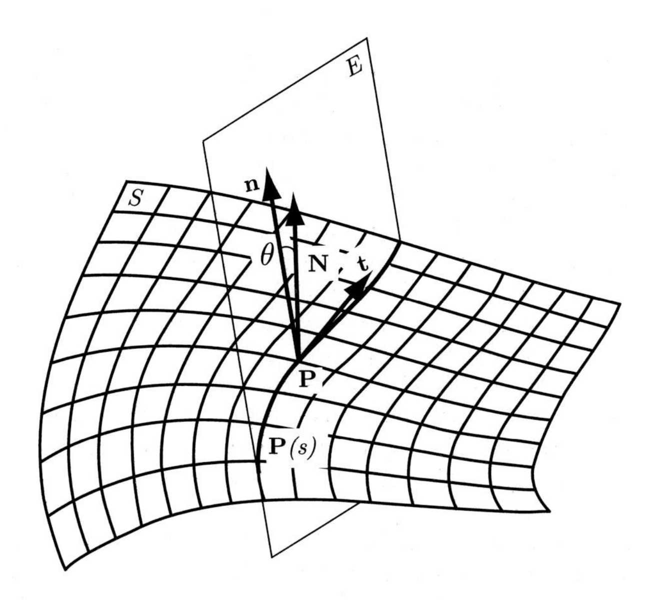

Figure 4.2: *Meusnier's Theorem.* Plane E intersects surface S to form curve $\mathbf{P}(s)$. $\mathbf{N}$ is the surface normal and $\mathbf{n}$ is the normal vector to $\mathbf{P}(s)$ and θ is the angle subtended by $\mathbf{N}$ and $\mathbf{n}$. The curvature of $\mathbf{P}(s)$ is $1/\rho$. Meusnier showed that the ratio $\cos\theta/\rho$ is constant so that $1/\rho$ reaches its maximum when $\theta = 0$; $\mathbf{P}(s)$ is then called a *normal section.*

In general, the unit surface normal, $\mathbf{N}$ and the unit normal to the curve, $\mathbf{n}$ are not equal. Refer to Fig. 4–1. Let θ be the angle subtended by these two vectors so that,

$$\mathbf{N}\cdot\mathbf{t}_s = \frac{\mathbf{N}\cdot\mathbf{n}}{\rho} = \frac{\cos\theta}{\rho} = \mathbf{N}\cdot\mathbf{P}_{vv}v_s^2 + 2\,\mathbf{N}\cdot\mathbf{P}_{vw}v_s w_s + \mathbf{N}\cdot\mathbf{P}_{ww}w_s^2, \tag{4.10}$$

where we use the fact that $\mathbf{N}\cdot\mathbf{P}_v = \mathbf{N}\cdot\mathbf{P}_w = 0$. Now we define the *second fundamental quantities*

$$L = \mathbf{N}\cdot\mathbf{P}_{vv}, \qquad M = \mathbf{N}\cdot\mathbf{P}_{vw}, \qquad N = \mathbf{N}\cdot\mathbf{P}_{ww}, \tag{4.11}$$

so that Eq. 4.10 can be written as

$$\frac{\cos\theta}{\rho} = Lv_s^2 + 2\,Mv_s w_s + Nw_s^2, \tag{4.12}$$

the right member of which is called the *second fundamental form.* The second fundamental quantities depend entirely on the properties of the surface and are therefore independent of the inscribed curve $\mathbf{P}$. The curve's unit tangent vector is $\mathbf{P}_s = \mathbf{t} = \mathbf{P}_v v_s + \mathbf{P}_w w_s$, this from Eq. 4.6.

Again refer to Fig. 4.2. Shown is a surface S, a point on that surface $\mathbf{P}$ and the unit normal vector $\mathbf{N}$ to the surface at $\mathbf{P}$. Through $\mathbf{P}$ there passes a plane E that cuts S in a curve that we will refer to as $\mathbf{P}(s)$. The curve is, of course, a plane curve; its unit normal vector $\mathbf{n}$ and its unit tangent vector $\mathbf{t}$ both lie on E. If we hold the direction of $\mathbf{t}$ fixed then, from Eq. 4.6, the coefficients v_s and w_s must also be fixed. It follows then that the entire right member of Eq. 4.12 is constant; this because the second fundamental quantities represent surface properties and are not related to the orientation of the vector $\mathbf{t}$.

Now rock the plane E as if it were hinged on the vector $\mathbf{t}$. The value of the right member of Eq. 4.12 remains constant. It follows that its left member also must be constant so that ρ must be proportional to $\cos\theta$. Our conclusion is that ρ attains its maximum value when θ is zero.

This is *Meusnier's Theorem.* When $\theta = 0$ and the normal $\mathbf{n}$ to the curve is in the same direction as the surface normal $\mathbf{N}$ then the radius of curvature of the arc attains its maximum value. The curvature of the curve $1/\rho$ is then at a minimum. Such a curve formed by the intersection of a surface and a plane that contains the surface normal is called a *normal section.*

4.4 The Theorem of Gauss

The mechanics of Gauss' theorem is shown in Fig. 4.3. Again we see the surface S, the point $\mathbf{P}$ and the unit normal vector $\mathbf{N}$. The difference now is that the plane E contains $\mathbf{N}$ so that the curve $\mathbf{P}(s)$ is now a normal section and its normal vector $\mathbf{n}$ coincides with $\mathbf{N}$. We rotate E about $\mathbf{N}$ just as a revolving door rotates about its axis.

Now we change the parameter back to t from s so that Eq. 4.12 becomes

$$\frac{1}{\rho} = \frac{Lv_t^2 + 2Mv_tw_t + Nw_t^2}{Ev_t^2 + 2Fv_tw_t + Gw_t^2}, \tag{4.13}$$

and Eq. 4.6 is again

$$\mathbf{t} = \frac{\mathbf{P}_v v_t + \mathbf{P}_w w_t}{\sqrt{Ev_t^2 + 2Fv_tw_t + Gw_t^2}}.$$

Our goal here is to find the extremum values of the curvature, $1/\rho$ as the plane pivots about the normal vector $\mathbf{N}$. We are to find those directions of $\mathbf{t}$ for which the curvature is a maximum or minimum. Now the direction of $\mathbf{t}$ is determined completely by the ratio $\lambda = w_t/v_t$ which, when substituted into Eqs. 4.6 and 4.13, results in

$$\mathbf{t} = \frac{\mathbf{P}_v + \mathbf{P}_w\lambda}{\sqrt{E + 2F\lambda + G\lambda^2}}, \tag{4.14}$$

and

$$\frac{1}{\rho} = \frac{L + 2M\lambda + N\lambda^2}{E + 2F\lambda + G\lambda^2}. \tag{4.15}$$

Note that if $\lambda = 0$ then $\mathbf{t}$ is the unit vector $\mathbf{P}_v/\sqrt{E}$ and when λ is infinite $\mathbf{t} = \mathbf{P}_w/\sqrt{G}$. It is clear that as λ assumes values over the interval $(-\infty, \infty)$ the vector $\mathbf{t}$ ranges over all directions perpendicular to $\mathbf{N}$; all other quantities in Eqs. 4.14 and 4.15 represent surface properties and are independent of the direction of $\mathbf{t}$.

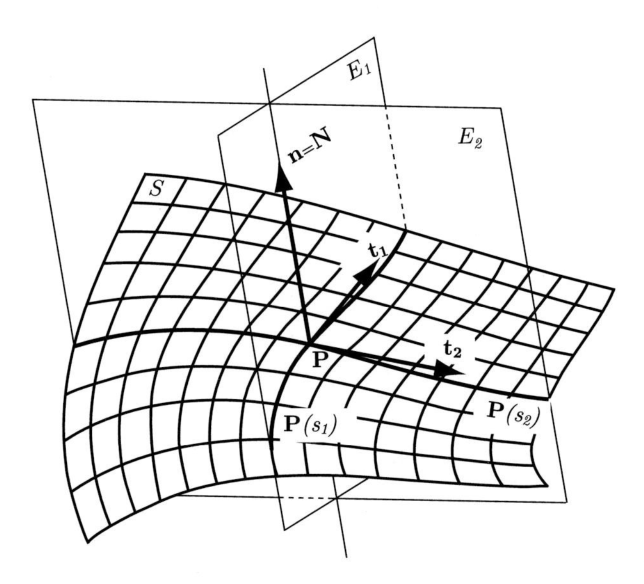

Figure 4.3: *The Theorem of Gauss.* $\mathbf{N}$ is the unit normal vector to the surface S at point $\mathbf{P}$. A plane E passes through $\mathbf{N}$ that intersects S in curve $\mathbf{P}(s)$ which is then a normal section. As E is rotated, pivoting on $\mathbf{N}$, a one-parameter family of normal sections is formed. Planes E_1 and E_2 are two positions of that plane and $\mathbf{P}(s_1)$ and $\mathbf{P}(s_2)$ are two of the resulting normal sections. Gauss showed that there are exactly two positions of E for which the curvature of $\mathbf{P}(s)$ is an extremum and that these two planes must be perpendicular. The two directions are called the *principal directions*. The corresponding curvatures are the *principal curvatures*.

At extremum values of curvature the first derivative of Eq. 4.15 must vanish so that

$$\frac{d}{d\lambda}\left(\frac{1}{\rho}\right) = \frac{2}{(E+2F\lambda+G\lambda^2)^2}\Big[(E+2F\lambda+G\lambda^2)(M+N\lambda) \\ -(L+2M\lambda+N\lambda^2)(F+G\lambda)\Big] = 0, \tag{4.16}$$

which reduces quickly to

$$(E+F\lambda)(M+N\lambda)-(L+M\lambda)(F+G\lambda)=0, \tag{4.17}$$

then to

$$\frac{1}{\rho} = \frac{M+N\lambda}{F+G\lambda} = \frac{L+M\lambda}{E+F\lambda}, \tag{4.18}$$

which comes from Eq. 4.15. Equation 4.16 also leads to the quadratic polynomial in λ

$$(EM-FL)+(EN-GL)\lambda+(FN-GM)\lambda^2=0, \tag{4.19}$$

the solution of which is

$$\lambda_{\pm} = \frac{-(EN-GL)\pm\mathcal{R}}{2(FN-GM)}, \tag{4.20}$$

where

$$\mathcal{R}^2 = (EN - GL)^2 - 4(EM - FL)(FN - GM)$$
$$= (EN + GL - 2FM)^2 - 4(EG - F^2)(LN - M^2). \tag{4.21}$$

By substituting the two values of λ from Eqs. 4.20 and 4.21 into Eq. 4.14 we get two direction vectors for which the curvature attains extremum values. They are unique to the surface, and are given by

$$\mathbf{t}_{\pm} = \frac{\mathbf{P}_v + \mathbf{P}_w \lambda_{\pm}}{\sqrt{E + 2F\lambda_{\pm} + G\lambda_{\pm}^2}}. \tag{4.22}$$

These are the two *principal directions*; one corresponds to the maximum value of curvature, the other to its minimum.

Consider next the two roots of the quadratic in Eq. 4.19, λ_+ and λ_-. Then, from another theorem by Gauss[3] that relates the roots of a polynomial to its coefficients, we get

$$\lambda_+ + \lambda_- = -\frac{EN - GL}{FN - GM}, \qquad \lambda_+ \lambda_- = \frac{EM - FL}{FN - GM}. \tag{4.23}$$

Now form the scalar product of the vectors in the two principal directions and get

$$\mathbf{t}_+ \cdot \mathbf{t}_- = \big[E + (\lambda_+ + \lambda_-)F + G\lambda_+\lambda_-\big]/H_+H_-. \tag{4.24}$$

On substituting for the sum and product of the two λ's, given in Eq. 4.22, we can see that $\mathbf{t}_+ \cdot \mathbf{t}_- = 0$ showing that the two principal directions must be orthogonal. This result is certainly not intuitively obvious.

The curvatures in the two principal directions are termed, naturally enough, the *principal curvatures*, obtained from Eqs. 4.15 and 4.18. There are two equivalent expressions for these principal curvatures. They are

$$\frac{1}{\rho_{\pm}} = \frac{M + N\lambda_{\pm}}{F + G\lambda_{\pm}} = \frac{L + M\lambda_{\pm}}{E + F\lambda_{\pm}}. \tag{4.25}$$

The reciprocals, $\rho_{\pm}$, are called the *principal radii of curvature* and are illustrated in Fig. 4.4.

4.5 Geodesics on a Surface

Choose any pair of points on a surface and connect them with all possible curves that can be inscribed on the surface. A *geodesic* curve is that curve whose arc length *on the surface* is an extremum. Clearly, to find geodesic curves we need to invoke the Calculus of Variations and the vector versions of the Euler equations derived in Chapter 1

$$\frac{d}{dt}\nabla_t f = \nabla f. \tag{1.26}$$

[3] Struik 1961, p. 80, Korn and Korn 1968, pp. 572–574.

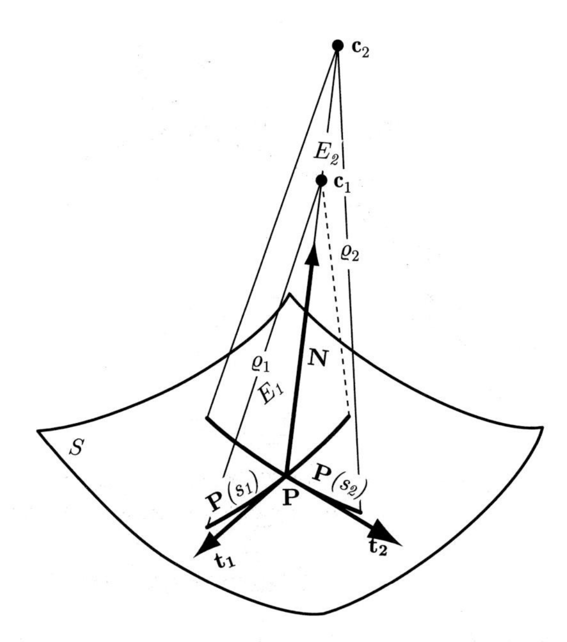

Figure 4.4: *Principal Radii of Curvature.* $\mathbf{N}$ is the unit normal vector to the surface S at point $\mathbf{P}$. Let planes E_1 and E_2 pass through $\mathbf{N}$ in the two principal directions and let $\mathbf{P}(s_1)$ and $\mathbf{P}(s_2)$ be the resulting normal sections. The two principal centers of curvature $\mathbf{c}_1$ and $\mathbf{c}_2$ lie on $\mathbf{N}$ and can be thought of as the centers of circles that best fit curves $\mathbf{P}(s_1)$ and $\mathbf{P}(s_2)$.

As usual f designates the variational integrand that is defined by the nature of the problem to be solved. In this case it is the length of the curves imbedded on the surface.

Let the surface be represented by a vector function of two variables, $\mathbf{P}(v,\ w)$. A curve etched on this surface will then be $\mathbf{P}(t) = \mathbf{P}\big(v(t),\ w(t)\big)$. We will use the derivative of its arc length given Eq. 4.5 in the variational integral given in Eq. 1.6

$$I = \int ds = \int \sqrt{Ev_t^2 + 2Fv_t w_t + Gw_t^2}\ dt. \tag{4.26}$$

The associated Euler equations are from Eq. 1.26

$$\begin{aligned} &\frac{d}{dt}\frac{\partial f}{\partial v_t} = \frac{\partial f}{\partial v}, \qquad \frac{d}{dt}\frac{\partial f}{\partial w_t} = \frac{\partial f}{\partial w}, \\ &f^2 = Ev_t^2 + 2Fv_t w_t + Gw_t^2 = \mathbf{P}_t^2 = (ds/dt)^2, \end{aligned} \tag{4.27}$$

the variational integrand, from Eq. 4.26. First we calculate the two partial derivatives

$$\begin{cases} f\dfrac{\partial f}{\partial v_t} = Ev_t + Fw_t = \mathbf{P}_v \cdot (\mathbf{P}_v v_t + \mathbf{P}_w w_t) = \mathbf{P}_v \cdot \mathbf{P}_t \\ f\dfrac{\partial f}{\partial v} = E_v v_t^2 + 2F_v v_t w_t + G_v w_t^2 \\ \quad = (\mathbf{P}_v \cdot \mathbf{P}_{vv}) v_t^2 + [(\mathbf{P}_{vv} \cdot \mathbf{P}_w) \\ \qquad + (\mathbf{P}_v \cdot \mathbf{P}_{vw})] v_t w_t + (\mathbf{P}_w \cdot \mathbf{P}_{vw}) w_t^2 \\ \quad = (\mathbf{P}_{vv} v_t + \mathbf{P}_{vw} w_t) \cdot (\mathbf{P}_v v_t + \mathbf{P}_w w_t) \\ \quad = \mathbf{P}_{vt} \cdot \mathbf{P}_t. \end{cases}$$

$$\frac{\partial f}{\partial v_t} = \frac{\mathbf{P}_v \cdot \mathbf{P}_t}{f} = \mathbf{P}_v \cdot \mathbf{P}_s \quad \text{and} \quad \frac{\partial f}{\partial v} = \frac{\mathbf{P}_{vt} \cdot \mathbf{P}_t}{f} = \mathbf{P}_{vt} \cdot \mathbf{P}_s. \tag{4.28}$$

In exactly the same way we find that

$$\frac{\partial f}{\partial w_t} = \mathbf{P}_w \cdot \mathbf{P}_s \quad \text{and} \quad \frac{\partial f}{\partial w} = \mathbf{P}_{wt} \cdot \mathbf{P}_s. \tag{4.29}$$

Now we come to the two Euler equations in Eq. 4.28. The first of these is

$$\frac{d}{dt}\frac{\partial f}{\partial v_t} - \frac{\partial f}{\partial v} = \frac{d}{dt}(\mathbf{P}_v \cdot \mathbf{P}_s) - (\mathbf{P}_{vt} \cdot \mathbf{P}_t) = \mathbf{P}_v \cdot \mathbf{P}_{ss} = 0,$$

with a similar expression for the second. Together they are

$$\mathbf{P}_v \cdot \mathbf{P}_{ss} = 0, \qquad \mathbf{P}_w \cdot \mathbf{P}_{ss} = 0, \tag{4.30}$$

the equations for a geodesic curve on the given surface. By multiplying the second of these by $\mathbf{P}_v$; the first by $\mathbf{P}_w$ and subtracting we get $\mathbf{P}_{ss} \times (\mathbf{P}_v \times \mathbf{P}_w) = D(\mathbf{P}_{ss} \times \mathbf{N}) = 0$. Since $\mathbf{P}_{ss} = \mathbf{t}_s = (1/\rho)\,\mathbf{n}$ it follows that $\mathbf{n} = \mathbf{N}$. Meusnier's theorem tells us that a geodesic must be a normal section.

4.6 The Weingarten Equations[4]

The definition of the second fundamental quantities given in Eq. 4.11 and the fact that $\mathbf{N} \cdot \mathbf{P}_v = \mathbf{N} \cdot \mathbf{P}_w = 0$ leads us to

$$\begin{cases} L = \mathbf{N} \cdot \mathbf{P}_{vv} = -\mathbf{N}_v \cdot \mathbf{P}_v \\ M = \mathbf{N} \cdot \mathbf{P}_{vw} = -\mathbf{N}_v \cdot \mathbf{P}_w = -\mathbf{N}_w \cdot \mathbf{P}_v \\ N = \mathbf{N} \cdot \mathbf{P}_{ww} = -\mathbf{N}_w \cdot \mathbf{P}_w. \end{cases} \tag{4.31}$$

[4]Dickson 1914, pp. 55-56, Korn and Korn 1968, pp. 55–56.

Since $\mathbf{N}$ is a unit vector the derivatives of $\mathbf{N}^2$ must vanish so that $\mathbf{N} \cdot \mathbf{N}_v = \mathbf{N} \cdot \mathbf{N}_w = 0$. This means that $\mathbf{N}_v$ and $\mathbf{N}_w$ must be perpendicular to $\mathbf{N}$ and therefore can be written as linear combinations of $\mathbf{P}_v$ and $\mathbf{P}_w$, thus

$$\begin{cases} \mathbf{N}_v = \alpha \mathbf{P}_v + \beta \mathbf{P}_w \\ \mathbf{N}_w = \mathbf{P}_v + \delta \mathbf{P}_w, \end{cases} \tag{4.32}$$

where $\alpha,\ \beta,\ \gamma$ and δ are to be determined. The scalar product of these two equations with $\mathbf{P}_v$ and $\mathbf{P}_w$ yield

$$\begin{aligned} \alpha E + \beta F &= -L, & \gamma E + \delta F &= -M, \\ \alpha F + \beta G &= -M, & \gamma F + \delta G &= -N. \end{aligned} \tag{4.33}$$

By solving these for the Greek letters and substituting the result back into the original equations in Eq. 4.32, we get

$$\begin{cases} \mathbf{N}_v = \dfrac{(FM - GL)\mathbf{P}_v + (FL - EM)\mathbf{P}_w}{EG - F^2} \\ \mathbf{N}_w = \dfrac{(FN - GM)\mathbf{P}_v + (FM - EN)\mathbf{P}_w}{EG - F^2}. \end{cases} \tag{4.34}$$

These are the Weingarten equations.

Now consider those parametric curves that are orthogonal; those for which $\mathbf{P}_v \cdot \mathbf{P}_w = F = 0$. Then the Weingarten equations specialize to

$$\begin{cases} \mathbf{N}_v = -\dfrac{L}{E}\mathbf{P}_v - \dfrac{M}{G}\mathbf{P}_w \\ \mathbf{N}_w = -\dfrac{M}{E}\mathbf{P}_v - \dfrac{N}{G}\mathbf{P}_w. \end{cases} \tag{4.35}$$

We will consider separately the two geodesic curves through a regular point on the surface; say, $\mathbf{P}(v,\ w_0)$ and $\mathbf{P}(v_0,\ w)$, which must be normal sections each being a member of a one-parameter family of geodesics. First look at the curve $\mathbf{P}(v,\ w_0)$ and introduce the arc length parameter by taking v as a function of s. The other curve, represented by $\mathbf{P}(s) = \mathbf{P}(v(s),\ w_0)$ so that the unit tangent vector in the v-direction must be ${}_v\mathbf{t} = \mathbf{P}_s = \mathbf{P}_v v_s$. But another way of writing ${}_v\mathbf{t}$ is as $\mathbf{P}_v/\sqrt{E}$. It follows that $v_s = 1/\sqrt{E}$ where $\sqrt{E}$ is a normalizing factor for $\mathbf{P}_v$. Using the definition in Eq. 2.15 we find the binormal vector associated with the v-curve to be ${}_v\mathbf{b} = {}_v\mathbf{t} \times \mathbf{n} = (\mathbf{P}_v/\sqrt{E}) \times \mathbf{N} = -\mathbf{P}_w/\sqrt{G}$. To do this recall that, from Eq. 4.7, $\mathbf{N} = (\mathbf{P}_v/\sqrt{E}) \times (\mathbf{P}_w/\sqrt{G})$.

Now recall the second of the Frenet-Serret equations from Chapter 2

$$\mathbf{n}_s = -\frac{1}{\rho}\mathbf{t} + \frac{1}{\tau}\mathbf{b} \tag{2.20}$$

and compare it with the first Weingarten equation for orthogonal parametric curves in Eq. 4.35

$$\begin{aligned}\mathbf{n}_s = \mathbf{N}_v v_s &= -\left(\frac{L}{\sqrt{E}}\frac{\mathbf{P}_v}{\sqrt{E}} + \frac{M}{\sqrt{G}}\frac{\mathbf{P}_w}{\sqrt{G})}\right)\frac{1}{\sqrt{E}} \\ &= -\left(\frac{L}{E}\,{}_v\mathbf{t} - \frac{M}{\sqrt{EG}}\,{}_v\mathbf{b}\right).\end{aligned} \tag{4.36}$$

From this comes

$$\frac{1}{\rho_v} = \frac{L}{E}, \qquad \frac{1}{\tau_v} = \frac{M}{\sqrt{EG}}. \tag{4.37}$$

We do the same thing for the parametric curve in the w-direction, given by $\mathbf{P}(s) = \mathbf{P}(v_0,\ w(s))$, so that $\mathbf{P}_s = \mathbf{P}_w\, w_s$. As before it turns out that $w_s = 1/\sqrt{G}$. Now ${}_w\mathbf{t} = \mathbf{P}_w/\sqrt{G}$ so that ${}_w\mathbf{b} = (\mathbf{P}_w/\sqrt{G}) \times \mathbf{N}$. But $\mathbf{P}_w/\sqrt{G} = {}_v\mathbf{b}$. It follows than that

$${}_w\mathbf{b} = {}_w\mathbf{t} \times \mathbf{N} = (\mathbf{P}_w/\sqrt{G}) \times \mathbf{N} = -{}_v\mathbf{b} \times \mathbf{N} = {}_v\mathbf{t} = \mathbf{P}_v/\sqrt{E}. \tag{4.38}$$

We return to the second Weingarten equation to get

$$\begin{aligned}\mathbf{n}_s = \mathbf{N}_w w_s &= -\left(\frac{M}{\sqrt{E}}\frac{\mathbf{P}_v}{\sqrt{E}} + \frac{N}{\sqrt{G}}\frac{\mathbf{P}_w}{\sqrt{G}}\right)\frac{1}{\sqrt{G}} \\ &= -\frac{N}{G}\,{}_w\mathbf{t} - \frac{M}{\sqrt{EG}}\,{}_w\mathbf{b},\end{aligned} \tag{4.39}$$

which gives us

$$\frac{1}{\rho_w} = \frac{N}{G}, \qquad \frac{1}{\tau_w} = -\frac{M}{\sqrt{EG}}. \tag{4.40}$$

An additional result, from Eqs. 4.37 and 4.40, is that $1/\tau_v$ and $1/\tau_w$ are equal in magnitude but opposite in sign. This permits us to define $1/\sigma$

$$\frac{1}{\sigma} = \frac{1}{\tau_v} = -\frac{1}{\tau_w} = \frac{M}{\sqrt{EG}}. \tag{4.41}$$

A matter of terminology: If $F = 0$ everywhere then the two families of parametric curves are *orthogonal*; through every regular point on the surface pass two parametric curves that are perpendicular. If $F = 0$ and the two orthogonal families of parametric curves are in principal directions they are called *conjugate*; through every regular point passes two curves, one from each family, that are not only orthogonal but have tangent vectors that are in the two principal directions at that point. We will see later on that for conjugate parametric curves $M = 0$.

4.7 Transformation of Parameters

As had been said many times before, a transformation of the variable parameters leaves the surface unchanged. We have already used a kind of parameter transformation in Eq. 4.2 when we defined a curve etched onto a surface. We do the same thing here but, instead of introducing a single curve on the surface we engrave on it a two parameter family of curves. To do this we introduce two new parameters, p and q, and set $v = v(p,\, q)$, and $w = w(p,\, q)$, so that the representation of the surface in terms of the new parameters is related to the old by

$$\mathbf{P}(p,\, q) = \mathbf{P}\big(v(p,\, q), w(p,\, q)\big). \tag{4.42}$$

So what we are doing here is simply a transformation of parameters quite like switching t and s in the discussion on space curves in Chapter 2.

The partial derivatives are

$$\begin{cases} \mathbf{P}_p = \mathbf{P}_v v_p + \mathbf{P}_w w_p \\ \mathbf{P}_q = \mathbf{P}_v v_q + \mathbf{P}_w w_q. \end{cases} \tag{4.43}$$

The transformation needs to be non singular so the Jacobian must not vanish

$$J = v_p\, w_q - v_q\, w_p = \begin{vmatrix} v_p & v_q \\ w_p & w_q \end{vmatrix} \neq 0. \tag{4.44}$$

From Eq. 4.4 we calculate the first fundamental quantities for the new system

$$\begin{cases} e = \mathbf{P}_p^2 & = E\, v_p^2 + 2F\, v_p\, w_p + G\, w_p^2 \\ f = \mathbf{P}_p \cdot \mathbf{P}_q & = E\, v_p\, v_q + F\,(v_p\, w_q + v_q\, w_p) + G\, w_p\, w_q \\ g = \mathbf{P}_q^2 & = E\, v_q^2 + 2F\, v_q\, w_q + G\, w_q^2. \end{cases} \tag{4.45}$$

We also take the transformed differential element of area d to be

$$d^2 = eg - f^2. \tag{4.46}$$

Equation 4.45 can be cast in matrix form

$$\begin{pmatrix} e & f \\ f & g \end{pmatrix} = \begin{pmatrix} v_p & w_p \\ v_q & w_q \end{pmatrix} \begin{pmatrix} E & F \\ F & G \end{pmatrix} \begin{pmatrix} v_p & v_q \\ w_p & w_q \end{pmatrix}. \tag{4.47}$$

By taking the determinant of both sides of this equation we can see that the differential element of area in the two systems are related by

$$d^2 = J^2 D^2, \tag{4.48}$$

where D is defined in Eq. 4.8. We can go further. Since $J \neq 0$ we may take inverses to get

$$J \begin{pmatrix} v_p & w_p \\ v_q & w_q \end{pmatrix}^{-1} = \begin{pmatrix} w_q & -w_p \\ -v_q & v_p \end{pmatrix} \quad \text{and} \quad J \begin{pmatrix} v_p & v_q \\ w_p & w_q \end{pmatrix}^{-1} = \begin{pmatrix} w_q & -v_q \\ -w_p & v_p \end{pmatrix},$$

so the inverse transform is

$$J^2 \begin{pmatrix} E & F \\ F & G \end{pmatrix} = \begin{pmatrix} w_q & -w_p \\ -v_q & v_p \end{pmatrix} \begin{pmatrix} e & f \\ f & g \end{pmatrix} \begin{pmatrix} w_q & -v_q \\ -w_p & v_p \end{pmatrix}, \tag{4.49}$$

which leads to

$$\begin{cases} J^2 E = e w_q^2 - 2 f w_p w_q + g w_p^2 \\ J^2 F = -e v_q w_q + f(v_p w_q + v_q w_p) - g v_p w_p \\ J^2 G = e v_q^2 - 2 f v_p v_q + g v_p^2. \end{cases} \tag{4.50}$$

The second derivatives of $\mathbf{P}$ are

$$\begin{cases} \mathbf{P}_{pp} = \mathbf{P}_{vv}\, v_p^2 + 2\,\mathbf{P}_{vw}\, v_p\, w_p + \mathbf{P}_{ww}\, w_p^2 + \mathbf{P}_v\, v_{pp} + \mathbf{P}_w w_{pp} \\ \mathbf{P}_{pq} = \mathbf{P}_{vv}\, v_p\, v_q + \mathbf{P}_{vw}\,(v_p\, w_q + v_q\, w_p) \\ \qquad\qquad + \mathbf{P}_{ww}\, w_p\, w_q + \mathbf{P}_v\, v_{pq} + \mathbf{P}_w w_{pq} \\ \mathbf{P}_{qq} = \mathbf{P}_{vv}\, v_q^2 + 2\,\mathbf{P}_{vw}\, v_q\, w_q + \mathbf{P}_{ww}\, w_q^2 + \mathbf{P}_v\, v_{q,q} + \mathbf{P}_w w_{qq}, \end{cases} \tag{4.51}$$

so that the new second fundamental quantities, calculated using Eq. 4.11, are

$$\begin{cases} l \;\; = L\, v_p^2 + 2M\, v_p\, w_p + N w_p^2 \\ m = L\, v_p\, v_q + M(v_p\, w_q + v_q\, w_p) + N\, w_p\, w_q \\ n \;\; = L v_q^2 + 2M\, v_q\, w_q + N\, w_q^2. \end{cases} \tag{4.52}$$

These also can be cast in matrix form as

$$\begin{pmatrix} l & m \\ m & n \end{pmatrix} = \begin{pmatrix} v_p & w_p \\ v_q & w_q \end{pmatrix} \begin{pmatrix} L & M \\ M & N \end{pmatrix} \begin{pmatrix} v_p & v_q \\ w_p & w_q \end{pmatrix}, \tag{4.53}$$

so that the inverse transform is

$$J^2 \begin{pmatrix} L & M \\ M & N \end{pmatrix} = \begin{pmatrix} w_q & -w_p \\ -v_q & v_p \end{pmatrix} \begin{pmatrix} l & m \\ m & n \end{pmatrix} \begin{pmatrix} w_q & -v_q \\ -w_p & v_p \end{pmatrix}, \tag{4.54}$$

resulting in

$$\begin{cases} J^2 L \;\; = \;\; l w_q^2 - 2\, m w_p w_q + n w_p^2 \\ J^2 M = -l v_q w_q + m(v_p w_q + v_q w_p) - n v_p w_p \\ J^2 N \;\; = \;\; l v_q^2 - 2\, m v_p v_q + n v_p^2. \end{cases} \tag{4.55}$$

4.8 When the Parametric Curves are Conjugates

Now let's suppose that each of these two sets of parametric curves are orthogonal so that $F = \mathbf{P}_v \cdot \mathbf{P}_w$ and $f = \mathbf{P}_p \cdot \mathbf{P}_q$ are both zero. Then let us assume further that the two principal directions are the p and q directions so that the p and q parametric curves are not only orthogonal but conjugate as well. Then we can set

$$w_p = \lambda_+\, v_p, \qquad w_q = \lambda_-\, v_q, \tag{4.56}$$

just as we had done in preparation for the calculations in Eq. 4.14. Here λ_+ and λ_- are the two roots of the quadratic equation in Eq. 4.19 that appear in Eqs. 4.20 and 4.21.

Substituting these into the expression for m, the second equation in Eq. 4.52, we get

$$m = v_p\, v_q[L + 2M(\lambda_+ + \lambda_-) + N\,\lambda_+\,\lambda_-], \tag{4.57}$$

which reduces to zero when the sums and products of the two λ's from Eq. 4.24 are used. This means that m is zero whenever the parametric curves are conjugates. This results in the following equations for the two sets of first fundamental quantities, from Eqs. 4.45 and 4.50

$$\begin{cases} e = Ev_p^2 + Gw_p^2 & J^2E = ew_q^2 + gw_p^2 \\ 0 = Ev_pv_q + Gw_pw_q & \quad 0 = ev_qw_q + gv_pw_p \\ g = Ev_q^2 + Gw_q^2 & J^2G = ev_q^2 + gv_p^2, \end{cases} \tag{4.58}$$

Eq. 4.52 becomes when $m = 0$

$$\begin{cases} l \;= L\,v_p^2 + 2M\,v_p\,w_p + Nw_p^2 \\ 0 = L\,v_p\,v_q + M(v_p\,w_q + v_q\,w_p) + N\,w_p\,w_q \\ n = Lv_q^2 + 2M\,v_q\,w_q + N\,w_q^2, \end{cases} \tag{4.59}$$

and Eq. 4.55 is

$$\begin{cases} J^2L \;= \;\; lw_q^2 + nw_p^2 \\ J^2M = -lv_qw_q - nv_pw_p \\ J^2N \;= \;\; lv_q^2 + nv_p^2. \end{cases} \tag{4.60}$$

Next suppose that the angle between $\mathbf{P}_p$ and $\mathbf{P}_v$ (and therefore between $\mathbf{P}_q$ and $\mathbf{P}_w$) is the angle θ. Then, from Eq. 4.43, and recalling that $F = f = 0$

$$\begin{cases} \mathbf{P}_p \cdot \mathbf{P}_v \;= Ev_p = \sqrt{eE}\,\cos\theta \\ \mathbf{P}_q \cdot \mathbf{P}_v \;= Ev_q = \sqrt{gE}\,\cos(\pi/2 + \theta) = -\sqrt{gE}\,\sin\theta \\ \mathbf{P}_p \cdot \mathbf{P}_w = Gw_p = \sqrt{eG}\,\cos(\pi/2 - \theta) = \;\sqrt{eG}\,\sin\theta \\ \mathbf{P}_q \cdot \mathbf{P}_w = Gw_q = \sqrt{gG}\,\cos\theta, \end{cases} \tag{4.61}$$

which enables us to obtain the following relations

$$\begin{aligned} v_p &= \;\;\sqrt{e/E}\,\cos\theta, & w_p &= \sqrt{e/G}\,\sin\theta, \\ v_q &= -\sqrt{g/E}\,\sin\theta, & w_q &= \sqrt{g/G}\,\cos\theta, \end{aligned} \tag{4.62}$$

which satisfy the equations in Eqs. 4.45 and 4.50. Here we have used the fact that $eg = J^2\,EG$ from Eq. 4.48.

The next step is to substitute the expressions in Eq. 4.62 into the equations in Eq. 4.59 to get

$$\begin{cases} \dfrac{l}{e} = \dfrac{L}{E}\cos^2\theta + \dfrac{N}{G}\sin^2\theta + 2\dfrac{M}{\sqrt{EG}}\sin\theta\cos\theta \\ 0 = -\left(\dfrac{L}{E} - \dfrac{N}{G}\right)\sin 2\theta + 2\dfrac{M}{\sqrt{EG}}\cos 2\theta \\ \dfrac{n}{g} = \dfrac{L}{E}\sin^2\theta + \dfrac{N}{G}\cos^2\theta - 2\dfrac{M}{\sqrt{EG}}\sin\theta\cos\theta, \end{cases} \tag{4.63}$$

to which we apply Eqs. 4.37, 4.40 and 4.41 to get

$$\begin{cases} \dfrac{1}{\rho_p} = \dfrac{\cos^2\theta}{\rho_v} + \dfrac{\sin^2\theta_w}{\rho_w} + \dfrac{2}{\sigma}\sin\theta\cos\theta \\ \dfrac{1}{\rho_q} = \dfrac{\sin^2\theta}{\rho_v} + \dfrac{\cos^2\theta_w}{\rho_w} - \dfrac{2}{\sigma}\sin\theta\cos\theta \\ \tan 2\theta = \dfrac{2/\sigma}{1/\rho_v - 1/\rho_w}. \end{cases} \tag{4.64}$$

Now we apply the relations in Eq. 4.62 to Eq. 4.60, again recalling that $f = F = 0$ and that $eg = J^2EG$ from Eq. 4.48 and get

$$\begin{cases} \dfrac{L}{E} = \dfrac{l}{e}\cos^2\theta + \dfrac{n}{g}\sin^2\theta \\ \dfrac{M}{\sqrt{EG}} = \left(\dfrac{l}{e} - \dfrac{n}{g}\right)\sin\theta\cos\theta \\ \dfrac{N}{G} = \dfrac{l}{e}\sin^2\theta + \dfrac{n}{g}\cos^2\theta, \end{cases} \tag{4.65}$$

which, from Eqs. 4.37, 4.40 and 4.41 become

$$\begin{cases} \dfrac{1}{\rho_v} = \dfrac{\cos^2\theta}{\rho_p} + \dfrac{\sin^2\theta}{\rho_q} \\ \dfrac{1}{\rho_w} = \dfrac{\sin^2\theta}{\rho_p} + \dfrac{\cos^2\theta}{\rho_q} \\ \dfrac{1}{\sigma} = \left(\dfrac{1}{\rho_p} - \dfrac{1}{\rho_q}\right)\sin\theta\cos\theta. \end{cases} \tag{4.66}$$

The equations in Eqs. 4.64 and 4.66 are reciprocal transforms. If $1/\rho_p$ and $1/\rho_q$ are the two principle curvatures then Eq. 4.66 provides $1/\rho_v$ and $1/\rho_w$, the curvatures of a geodesic curve whose tangent makes an angle θ with one of the principal directions. The torsion of that curve is then $1/\sigma$. These relations will be most important in subsequent chapters.

4.9 When F $\neq$ 0

This is the more general case when the given system of parametric curves are not orthogonal but the transformation is to a system in which these curves are conjugate. Refer back to Eqs. 4.45 and 4.52 and set $m = f = 0$ to get

$$\begin{cases} E\, v_p\, v_q + F\,(v_p\, w_q + v_q\, w_p) + G\, w_p\, w_q = 0 \\ L\, v_p\, v_q + M\,(v_p\, w_q + v_q\, w_p) + N\, w_p\, w_q = 0. \end{cases} \tag{4.67}$$

Once again $\mathbf{P}_p/\sqrt{e}$ and $\mathbf{P}_q/\sqrt{g}$ are unit vectors in the two principal directions so that we may invoke Eq. 4.56 and Eq. 4.67 becomes

$$\begin{cases} E + F\,(\lambda_+ + \lambda_-) + G\lambda_+\lambda_- = 0 \\ L + M\,(\lambda_+ + \lambda_-) + N\lambda_+\lambda_- = 0. \end{cases} \tag{4.68}$$

When we apply Eq. 4.24 we can see that both are satisfied.

The remaining equations of Eq. 4.45 become, using this notation

$$\begin{cases} e = v_p^2(E + 2F\lambda_+ + G\lambda_+^2) \\ g = v_q^2(E + 2F\lambda_- + G\lambda_-^2) \end{cases} \tag{4.69}$$

and those of Eq. 4.52 are

$$\begin{cases} l\ = v_p^2(L + 2M\lambda_+ + N\lambda_+^2) \\ n = v_q^2(L + 2M\lambda_- + N\lambda_-^2). \end{cases} \tag{4.70}$$

In terms of the p, q-parameters, which correspond to a conjugate system of parametric curves, we may write the principal directions as unit vectors

$$\mathbf{T_p} = \mathbf{P}_p/\sqrt{e}, \qquad \mathbf{T_q} = \mathbf{P}_q/\sqrt{g}. \tag{4.71}$$

Also, from Eqs. 4.37, 4.40 and 4.41, we may write the principal curvatures as

$$1/\rho_p = l/e, \qquad 1/\rho_q = n/g. \tag{4.72}$$

So now we can start with a surface described in any system of parameters, calculate the surface normal and the first and second fundamental quantities and then transform to a new parametric system in which the lines of curvature are conjugate by using Eqs. 4.69 and 4.70. We will use these results in Chapter 9.

Now is a good time to introduce a few new concepts. First, *asymptotic directions* and *asymptotic curves*. Suppose the second fundamental form, Eq. 4.12, is *zero* yielding a total differential equation

$$Lv_t^2 + 2Mv_t w_t + Nw_t^2 = 0, \tag{4.73}$$

whose solution is an expression for an arc on the surface with the property that its curvature is *zero*. These are the *asymptotic curves* or *asymptotic lines*. Any straight line embedded on the surface, as it turns out, is asymptotic.

A second new concept is a little strange. Suppose the first fundamental form vanishes, resulting in a differential equation for another curve embedded on the surface

$$E\,v_t^2 + 2F\,v_t\,w_t + G\,w_t^2 = 0. \tag{4.74}$$

Such a curve is called an *isotropic curve*. But from Eq. 4.5 this means that the arc length along such a curve must always be *zero* which is impossible for any real curve; for this reason it is sometimes referred to as a *null curve*. An isotropic curve, if it exists, must be complex which opens the door to some amusing speculations especially in an optical context.

Our interest here is in the way these curves, asymptotic curves and isotropic curves, can be used to understand the geometric properties of surfaces and their ultimate application to the structure of wavefronts and the geometric aberrations associated with an optical system. These will be treated further in Chapter 6 and applied in Chapter 8.

4.10 The Structure of the Prolate Spheroid[5]

By rotating a conic section about one of its axes we generate a conic section of revolution. If the axis of revolution is the major axis of the ellipse the surface generated is called a *prolate spheroid*; if it is the minor axis, the surface is an *oblate spheroid*. Here we will deal with only the prolate spheroid.

As in the case of the planar conic section described in Chapter 2 we choose a focus as the pole and the polar axis as the axis of revolution and use the polar coordinates ϕ, θ and ϱ. If ϵ is the eccentricity of the generating conic and if r is its semi latus rectum, then the surface is represented, using a vector notation, by

$$\mathbf{P} = \varrho(\sin\phi\sin\theta,\ \sin\phi\cos\theta,\ \cos\phi), \tag{4.75}$$

where ϱ is taken from Chapter 2

$$\varrho = \frac{r}{1-\epsilon\cos\phi}. \tag{2.33}$$

Clearly θ is the parameter of the rotation that generates the surface.

In this way we can describe four conic surfaces of revolution; the sphere $\epsilon = 0$, the prolate spheroid $\epsilon < 1$, the paraboloid of revolution $\epsilon = 1$ and the hyperboloid of two sheets $\epsilon > 1$. In what follows we will concentrate on the prolate spheroid although the results that we will obtain may be extended easily to the other three types of conic surfaces. As before we refer to the focus at the pole as the spheroid's *proximal focus*; the other is its *distal focus*.

First note that the derivative of ϱ with respect to ϕ is

$$\varrho_\phi = \frac{-\epsilon\sin\phi}{1-\epsilon\cos\phi}\varrho. \tag{4.76}$$

Then it can be shown that

$$\begin{cases} \mathbf{P}_\phi = \dfrac{\varrho}{1-\epsilon\cos\phi}\Big((\cos\phi-\epsilon)\sin\theta, (\cos\phi-\epsilon)\cos\theta, -\sin\phi\Big) \\ \mathbf{P}_\theta = \varrho\sin\phi\big(\cos\theta,\ -\sin\theta,\ 0\big), \end{cases} \tag{4.77}$$

[5]Struik 1961, pp. 133–136. Clegg 1968, pp. 149–152, Bolza 1961, pp. 164–166.

so that the first fundamental quantities are, from Eq. 4.4

$$\begin{cases} E = \mathbf{P}_\phi^2 & = \dfrac{\varrho^2}{(1-\epsilon\cos\phi)^2}\mathcal{K}^2 \\ F = \mathbf{P}_\phi \cdot \mathbf{P}_\theta = 0 & \\ G = \mathbf{P}_\theta^2 & = \varrho^2 \sin^2\phi, \end{cases} \tag{4.78}$$

where from Chapter 2

$$\mathcal{K}^2 = 1+\epsilon^2-2\epsilon\cos\phi. \tag{2.39}$$

Since $F = 0$ the parametric curves must be orthogonal. From Eq. 4.8

$$D^2 = EG - F^2 = \frac{\varrho^4 \sin^2\phi}{(1-\epsilon\cos\phi)^2}\mathcal{K}^2. \tag{4.79}$$

From Eq. 4.1 the surface normal is the vector product of these two vectors

$$\mathbf{P}_\phi \times \mathbf{P}_\theta = \frac{-\varrho^2 \sin\phi}{1-\epsilon\cos\phi}\big(\sin\phi\sin\theta,\ \sin\phi\cos\theta,\ \cos\phi - \epsilon\big). \tag{4.80}$$

Division by D results in the unit normal vector to the surface

$$\mathbf{N} = -\frac{1}{\mathcal{K}}\big(\sin\phi\sin\theta,\ \sin\phi\cos\theta,\ \cos\phi - \epsilon\big). \tag{4.81}$$

To calculate the second fundamental quantities we need the derivatives of N. These are

$$\begin{cases} \mathbf{N}_\phi = -\dfrac{1}{\mathcal{K}}\big(\cos\phi\sin\theta,\ \cos\phi\cos\theta,\ -\sin\phi\big) + \dfrac{\mathcal{K}_\phi}{\mathcal{K}^2}\mathbf{N} \\ \mathbf{N}_\theta = -\dfrac{\sin\phi}{\mathcal{K}}\big(\cos\theta,\ -\sin\theta,\ 0\big), \end{cases} \tag{4.82}$$

and the second fundamental quantities are, since $\mathbf{P}_\phi \cdot \mathbf{N} = \mathbf{P}_\theta \cdot \mathbf{N} = 0$

$$\begin{cases} L = -\mathbf{N}_\phi \cdot \mathbf{P}_\phi = \dfrac{\varrho}{\mathcal{K}} \\ M = -\mathbf{N}_\phi \cdot \mathbf{P}_\theta = -\mathbf{N}_\theta \cdot \mathbf{P}_\phi = 0 \\ N = -\mathbf{N}_\theta \cdot \mathbf{P}_\theta = \dfrac{\varrho \sin^2\phi}{\mathcal{K}}. \end{cases} \tag{4.83}$$

Here we have used Eqs. 4.11 and 4.32. We know from this that the parametric curves are conjugates since $M = 0$.

The first task is to substitute for the first and second fundamental quantities in the quadratic equation from Eq. 4.19 whose solutions are found in Eqs. 4.20 and 4.21. Since both $F = 0$

and $M = 0$, Eq. 4.19 is ephemeral and we may take as its roots $\lambda_+ = 0$ and $\lambda_- = \infty$. The principal directions given as unit vectors, from Eq. 4.22, are then

$$\begin{cases} \mathbf{T}_+ = \dfrac{\mathbf{P}_\phi}{\sqrt{E}} = \dfrac{1}{\mathcal{K}}\big((\cos\phi - \epsilon)\sin\theta,\ (\cos\phi - \epsilon)\cos\theta,\ -\sin\phi\big) \\ \mathbf{T}_- = \dfrac{\mathbf{P}_\theta}{\sqrt{G}} = (\cos\theta,\ -\sin\theta,\ 0). \end{cases} \tag{4.84}$$

Now we come to the calculation of the principal curvatures for which we use Eq. 4.25

$$\begin{cases} \dfrac{1}{\rho_+} = \dfrac{L}{E} = \dfrac{(1-\epsilon\cos\phi)^2}{\varrho\,\mathcal{K}^3} \\ \dfrac{1}{\rho_-} = \dfrac{N}{G} = \dfrac{1}{\varrho\,\mathcal{K}}, \end{cases} \tag{4.85}$$

from which comes, using Eq. 2.43

$$\begin{cases} \rho_+ = \dfrac{r\mathcal{K}^3}{(1-\epsilon\cos\phi)^3} \\ \rho_- = \dfrac{r\mathcal{K}}{1-\epsilon\cos\phi}. \end{cases} \tag{4.86}$$

Next we find the principal *centers* of curvature. We start at $\mathbf{P}$ and move a distance $\rho_\pm$ along the unit normal vector $\mathbf{N}$ to get $\mathbf{D}_\pm$ using the equation

$$\mathbf{D}_\pm = \mathbf{P} + \rho_\pm \mathbf{N}. \tag{4.87}$$

We calculate first $\mathbf{D}_-$ using Eqs. 4.71, 4.76 and 2.42 as well as the expression for ρ from Eq. 2.27 and get

$$\mathbf{D}_- = \frac{r\epsilon}{1-\epsilon\cos\phi}\mathbf{Z} = \epsilon\varrho\mathbf{Z}, \tag{4.88}$$

where $\mathbf{Z}$ is the unit vector along the positive polar axis.

The other principal center of curvature $\mathbf{D}_+$ is more complicated. To make life easier first note that $\mathbf{N} = (-1/\mathcal{K})\big[(1/\varrho)\mathbf{P} - \epsilon\mathbf{Z}\big]$ this from Eq. 4.81. Then from Eq. 4.87 we can get

$$\begin{aligned} \mathbf{D}_+ &= \mathbf{P} - \left(\frac{\varrho\mathcal{K}^2}{(1-\cos\phi)^2}\right)\left(\frac{1}{\varrho}\mathbf{P} - \epsilon\mathbf{Z}\right) \\ &= \frac{1}{(1-\epsilon\cos\phi)^2}\left\{\big[(1-\epsilon\cos\phi)^2 - \mathcal{K}^2\big]\mathbf{P} + \epsilon\varrho\mathcal{K}^2\mathbf{Z}\right\} \\ &= \frac{1}{(1-\epsilon\cos\phi)^2}\big[\mathcal{K}^2\,\mathbf{D}_- - \epsilon^2\sin^2\phi\,\mathbf{P}\big], \end{aligned} \tag{4.89}$$

where $\mathcal{K}$ is given in Eq. 2.39 and $\mathbf{P}$, in Eq. 4.75. These results will be used in Chapters 7 and 13.

Notice that one of these points $\mathbf{D}_-$ lies on the z-axis, the axis of rotational symmetry, just as one would expect for a rotationally symmetric surface. The surface $\mathbf{D}$ is, of course, the two-surfaced *evolute* of the prolate spheroid.

4.11 Other Ways of Representing Surfaces

In this chapter we have concentrated on surfaces that are represented by functions of two parameters. Now is the time to consider alternatives. We can represent surfaces explicitly, where one of the coordinates is given as a function of the other two, or implicitly, where the surface is given as a function of all three coordinates that is set equal to a constant.

In the first instance suppose that z is given as a function of x and y, $z = z(x, y)$. We can use the parametric representation of the surface where the parameters are now x and y so that so that we can use Eq. 4.1 to calculate the surface normal. If we revert to the vector notation we have

$$\mathbf{P}(x, y) = \big(x, y, z(x, y)\big) \tag{4.90}$$

so that the two partial derivatives are then

$$\begin{cases} \mathbf{P}_x = \big(1, 0, z_x\big) \\ \mathbf{P}_y = \big(0, 1, z_y\big), \end{cases} \tag{4.91}$$

and their vector product is

$$\mathbf{P}_x \times \mathbf{P}_y = -\big(z_x, z_y, -1\big) \tag{4.92}$$

which must be a surface normal. The *unit* normal vector must then be

$$\mathbf{N} = -\frac{1}{\sqrt{1 + z_x^2 + z_y^2}}\big(z_x, z_y, -1\big). \tag{4.93}$$

The other possibility is that the surface is defined by a function of three coordinates

$$f\big(x, y, z\big) = constant. \tag{4.94}$$

We can interpret this equation as defining implicitly the function $z = z(x, y)$ so that the total derivatives are

$$\begin{cases} \dfrac{df}{dx} \equiv \dfrac{\partial f}{\partial x} + \dfrac{\partial f}{\partial z}\dfrac{\partial z}{\partial x} = 0 \\ \dfrac{df}{dy} \equiv \dfrac{\partial f}{\partial y} + \dfrac{\partial f}{\partial z}\dfrac{\partial z}{\partial y} = 0. \end{cases} \tag{4.95}$$

Reverting to the subscript notation for derivatives we can see that this is equivalent to

$$\begin{cases} z_x = -f_x/f_z \\ z_y = -f_y/f_z. \end{cases} \tag{4.96}$$

When these are substituted back into Eq. 4.93 we get

$$\mathbf{N} = \frac{1}{\sqrt{f_x^2 + f_y^2 + f_z^2}}\big(f_x, f_y, f_z\big) \tag{4.97}$$

which is exactly the normalized gradient of f

$$\mathbf{N} = \frac{\nabla f}{\sqrt{(\nabla f)^2}}, \tag{4.98}$$

just as it should be.

This concludes the chapter on the Differential Geometry of Surfaces in which we derived some of the general characteristics of surfaces represented by a vector function of two parameters. These include principal curvatures and principal directions, the properties of embedded geodesic curves, the Weingarten equations and the transformation of parameters. As an example, we looked at the conic section of revolution to which we will refer later. The treatment is certainly not complete but sufficient for out needs in subsequent chapters.

5 Partial Differential Equations of the First Order

In Chapter 3 we derived the *eikonal equation* (Eq. 3.65) from Fermat's principle of least time. The Hilbert Integral from the Calculus of Variations was introduced which led to the Hamilton-Jacobi theory of image formation. We also derived the eikonal equation for homogeneous, isotropic media and obtained its general solution which will be the subject of this chapter.

The vector form of the eikonal equation as derived previously is

$$(\nabla\phi)^2 = n^2, \tag{3.65}$$

which can be expressed in scalar form as

$$\left(\frac{\partial\phi}{\partial x}\right)^2 + \left(\frac{\partial\phi}{\partial y}\right)^2 + \left(\frac{\partial\phi}{\partial z}\right)^2 = n^2, \tag{5.1}$$

clearly a non linear first order partial differential equation.

As always we will begin with simpler problems, then generalize to the more difficult. Here we will begin with linear first order partial differential equations, with as much generality as is appropriate, using the method of Lagrange, [1] then proceed to specialize our results to non linear equations by applying the method of Lagrange and Charpit as extended by Jacobi. [2] We will develop general solutions for both linear and non linear equations. Using these methods we will then find solutions to Eq. 5.1 and apply them to geometrical optics. We will not consider methods other than these nor will we look at boundary value problems.

Just as general solutions of first order *ordinary* differential equations involve no more than one constant of integration, solutions of first order *partial* differential equations contain no more than one arbitrary function, the number of parameters of which is *one* less than the number of independent variables in the differential equation. We will use this fact later when we derive a general solution from a complete integral.

Recall that the eikonal is the optical path length from some arbitrary, fixed object point, through the optical system, to some point in image space. The coordinates of that point in image space are the arguments of the eikonal function. The coordinates of the point in object space do not enter these calculations and shall be ignored.

Setting the general solution of the eikonal equation equal to a constant yields an expression for the locus of points that have equal optical path lengths to the object point; in other words, a *surface of constant phase* or a *wavefront*.

---

[1] Forsyth 1959, Vol. V Chapter I. Garabedian 1998, pp. 6–16.

[2] Forsyth 1959, Vol. V, pp. 55–89 Chapters VI and VIII. Forsyth 1996, pp. 392–407. Garabedian 1998, pp. 18–24. Cohen 1933, pp. 250–253.

The Mathematics of Geometrical and Physical Optics: The k-function and its Ramifications. O.N. Stavroudis

ISBN: 3-527-40448-1

To get this general solution we need to go through a number of steps, each of which will be examined and explained as we proceed. Many of these steps will prove to be useful in subsequent chapters.

We have already encountered the total differential equation in Chapter 3. Here we will first look at the linear first order partial differential equation, then for the non linear case we will look at the bilinear concomitant which will lead us to the method of Lagrange and Charpit as modified by Jacobi. This leads to a solution called the *complete integral* which we then convert to the *general solution.*

This we will apply to the eikonal equation. If we set the general solution equal to an optical path length we introduce yet another parameter s representing the geometric distance between successive wavefronts and with this additional parameter we get an expression for a wavefront train.

5.1 The Linear Equation. The Method of Characteristics

This is also known as the *method of Lagrange*; [3] it has to do with equations in the form

$$P\frac{\partial \phi}{\partial x} + Q\frac{\partial \phi}{\partial y} + R\frac{\partial \phi}{\partial z} = S, \tag{5.2}$$

linear, first order partial differential equations. (Of course such an equation can contain any number of independent variables but for the sake of convenience here we use only three.)

If $S = 0$ the equation is called *homogeneous*; otherwise, it is *inhomogeneous*. Consider the homogeneous case first

$$P\frac{\partial \phi}{\partial x} + Q\frac{\partial \phi}{\partial y} + R\frac{\partial \phi}{\partial z} = 0, \tag{5.3}$$

and assume that $\phi(x,\ y,\ z) = 0$ is a solution. By comparing Eq. 5.3 with the total differential of ϕ

$$d\phi = \frac{\partial \phi}{\partial x}dx + \frac{\partial \phi}{\partial y}dy + \frac{\partial \phi}{\partial z}dz = 0. \tag{5.4}$$

we can see that they are proportional; that there exists a function μ such that

$$dx = Pd\mu \qquad dy = Qd\mu \qquad dz = Rd\mu. \tag{5.5}$$

By eliminating μ we get the *characteristic equations*

$$\frac{dx}{P} = \frac{dy}{Q} = \frac{dz}{R}. \tag{5.6}$$

Equation 5.6 can be generalized to equations involving any number of independent variables.

Equation 5.6 consists of no more than two total differential equations. Generally, the number of total differential equations is *one* less that the number of independent variables.

[3] Cohen 1933, pp. 15, 34–36. Korn and Korn 1968, pp. 101–102.

Each has solutions; these must be functionally independent; between them there can be no functional relationship. (In the more general case no characteristic can be a function of any of the others.) These solutions are called *characteristic functions* or, more simply, *characteristics*. The general solution of Eq. 5.2 is an arbitrary function of these two characteristics.

A very useful trick for finding characteristics comes from Eq. 5.5. If there exists functions $\alpha,\ \beta,$ and γ such that

$$\alpha P + \beta Q + \gamma R = 0 \tag{5.7}$$

then

$$\alpha dx + \beta dy + \gamma dz = 0. \tag{5.8}$$

These two relationships lead to a method for forming linear combinations of the characteristic equations in order to simplify the process of integration.

Consider the following trivial example

$$x\frac{\partial \phi}{\partial x} + y\frac{\partial \phi}{\partial y} + z\frac{\partial \phi}{\partial z} = 0.$$

The characteristic equations are

$$\frac{dx}{x} = \frac{dy}{y} = \frac{dz}{z},$$

which yields the two independent total differential equations

$$ydx - xdy = 0, \qquad zdy - ydz = 0,$$

the solutions of which are the two characteristics

$$\alpha = y/x, \qquad \beta = z/y,$$

so that the general solution is

$$\phi = \phi(\alpha,\ \beta) = \phi(y/x, z/y).$$

Calculating the partial derivatives of ϕ and plugging them into the original differential equation shows that it is indeed a solution.

Now for the inhomogeneous case

$$P\frac{\partial \phi}{\partial x} + Q\frac{\partial \phi}{\partial y} + R\frac{\partial \phi}{\partial z} = S.$$

Suppose the solution is an implicit function, $\psi(\phi,\ x,\ y,\ z)\ = \text{constant}.$ Then its partial derivatives are

$$\frac{\partial \psi}{\partial x} + \frac{\partial \psi}{\partial \phi}\frac{\partial \phi}{\partial x} = 0, \qquad \frac{\partial \psi}{\partial y} + \frac{\partial \psi}{\partial \phi}\frac{\partial \phi}{\partial y} = 0, \qquad \frac{\partial \psi}{\partial z} + \frac{\partial \psi}{\partial \phi}\frac{\partial \phi}{\partial z} = 0, \tag{5.9}$$

which gives us

$$\frac{\partial \phi}{\partial x} = -\frac{\partial \psi}{\partial x} \Big/ \frac{\partial \psi}{\partial \phi}, \qquad \frac{\partial \phi}{\partial y} = -\frac{\partial \psi}{\partial y} \Big/ \frac{\partial \psi}{\partial \phi}, \qquad \frac{\partial \phi}{\partial z} = -\frac{\partial \psi}{\partial z} \Big/ \frac{\partial \psi}{\partial \phi}, \tag{5.10}$$

which, when inserted into the original equation yields

$$P\frac{\partial \psi}{\partial x} + Q\frac{\partial \psi}{\partial y} + R\frac{\partial \psi}{\partial z} + S\frac{\partial \psi}{\partial \phi} = 0, \tag{5.11}$$

a homogeneous equation in four independent variables. Thus the problem is reduced to one already solved.

5.2 The Homogeneous Function

The word *homogeneous* crops up, confusingly, in a number of distinctly different contexts; as a *linear homogeneous partial differential equation*, as a *homogeneous optical medium* and also as the *homogeneous function* that we encountered in Chapter 1.

From Eq. 1.19 recall that a function $f(x,\ y,\ z)$ is said to be homogeneous if, for every constant λ, $f(\lambda x,\ \lambda y,\ \lambda z) = \lambda\, f(x,\ y,\ z)$. The derivative of this with respect to λ is

$$x\frac{\partial f}{\partial x} + y\frac{\partial f}{\partial y} + z\frac{\partial f}{\partial z} = f, \tag{5.12}$$

an inhomogeneous partial differential equation to which we may apply the results of the previous section, in particular Eq. 5.11, which we use to get the characteristic equations

$$\frac{dx}{x} = \frac{dy}{y} = \frac{dz}{z} = \frac{df}{f} \tag{5.13}$$

which leads us to the three total differential equations

$$z\,dx - x\,dz = 0, \qquad z\,dy - y\,dz = 0, \qquad z\,df - f\,dz = 0, \tag{5.14}$$

that lead to the three characteristics

$$\alpha = x/z, \qquad \beta = y/z, \qquad \gamma = f/z, \tag{5.15}$$

so that the general solution must be

$$\mathcal{F}(\alpha,\ \beta,\ \gamma) = \mathcal{F}(x/z,\ y/z,\ f/z) = \text{constant}, \tag{5.16}$$

in which f is defined implicitly. In principle we can solve this for f and get

$$f = z\,\mathcal{G}(x/z,\ y/z), \tag{5.17}$$

where $\mathcal{G}$ is another arbitrary function. This indeed does satisfy Eq. 5.12 and it also satisfies the definition of the homogeneous function.

5.3 The Bilinear Concomitant

For the moment, only as a matter of convenience, let us restrict ourselves to partial differential equations with only two independent variables. Let z be the dependent variable and let $p = \partial z/\partial x$ and $q = \partial z/\partial y$ be its partial derivatives. What follows can easily be generalized to any countable number of independent variables.[4]

Suppose we have a non linear, first order partial differential equation in the form

$$\mathcal{F}(x,\ y,\ z,\ p,\ q) = 0, \tag{5.18}$$

for which we seek a solution. Its helpful to think of $\mathcal{F}$ as a function of the five independent variables $x,\ y,\ z,\ p$ and q, forgetting for the moment that p and q are themselves partial derivatives of z.

Suppose further that there is a second differential equation

$$\mathcal{G}(x,\ y,\ z,\ p,\ q) = 0, \tag{5.19}$$

that possesses the same solution. Then Eqs. 5.19 and 5.20 can be considered as a simultaneous pair and solved for p and q. However, $\mathcal{F}$ and $\mathcal{G}$ must be functionally independent; the Jacobian must not vanish

$$\frac{\partial(\mathcal{F},\mathcal{G})}{\partial(p,q)} \neq 0. \tag{5.20}$$

When Eqs. 5.18 and 5.19 are solved for p and q they will be functions of $x,\ y,$ and z. These are then substituted into the total differential equation

$$dz = p\,dx + q\,dy, \tag{5.21}$$

which is then solved. As we have seen previously in Eq. 3.25 the condition for it to be exact is

$$\frac{\partial p}{\partial y} = \frac{\partial q}{\partial x}. \tag{5.22}$$

Now we take the derivatives of Eqs. 5.18 and 5.19 to get

$$\begin{cases} \sigma\frac{d\mathcal{F}}{dx} = \dfrac{\partial \mathcal{F}}{\partial x} + \dfrac{\partial \mathcal{F}}{\partial z}p + \dfrac{\partial \mathcal{F}}{\partial p}p_x + \dfrac{\partial \mathcal{F}}{\partial q}q_x = 0 \\ \dfrac{d\mathcal{G}}{dx} = \dfrac{\partial \mathcal{G}}{\partial x} + \dfrac{\partial \mathcal{G}}{\partial z}p + \dfrac{\partial \mathcal{G}}{\partial p}p_x + \dfrac{\partial \mathcal{G}}{\partial q}q_x = 0 \\ \dfrac{d\mathcal{F}}{dy} = \dfrac{\partial \mathcal{F}}{\partial y} + \dfrac{\partial \mathcal{F}}{\partial z}p + \dfrac{\partial \mathcal{F}}{\partial p}p_y + \dfrac{\partial \mathcal{F}}{\partial q}q_y = 0 \\ \dfrac{d\mathcal{G}}{dy} = \dfrac{\partial \mathcal{G}}{\partial y} + \dfrac{\partial \mathcal{G}}{\partial z}q + \dfrac{\partial \mathcal{G}}{\partial p}p_y + \dfrac{\partial \mathcal{G}}{\partial q}q_y = 0. \end{cases} \tag{5.23}$$

[4]Forsyth 1996, pp. 420–423. Cohen 1933, pp. 264–265.

Now multiply the first of these by $\partial\mathcal{G}/\partial p$ and the second, by $\partial\mathcal{F}/\partial q$ to eliminate p_x and get

$$\left(\frac{\partial\mathcal{F}}{\partial x}\frac{\partial\mathcal{G}}{\partial p}-\frac{\partial\mathcal{G}}{\partial x}\frac{\partial\mathcal{F}}{\partial p}\right)+p\left(\frac{\partial\mathcal{F}}{\partial z}\frac{\partial\mathcal{G}}{\partial p}-\frac{\partial\mathcal{G}}{\partial z}\frac{\partial\mathcal{F}}{\partial p}\right)$$
$$+q_x\left(\frac{\partial\mathcal{F}}{\partial q}\frac{\partial\mathcal{G}}{\partial p}-\frac{\partial\mathcal{G}}{\partial q}\frac{\partial\mathcal{F}}{\partial p}\right)=0,$$

which can be written in a more compact form using the Jacobian notation

$$\frac{\partial(\mathcal{F},\mathcal{G})}{\partial(x,p)}+p\frac{\partial(\mathcal{F},\mathcal{G})}{\partial(z,p)}+q_x\frac{\partial(\mathcal{F},\mathcal{G})}{\partial(q,p)}=0. \tag{5.24}$$

In exactly the same way, from the third and fourth equations of Eq. 5.23, we can get

$$\frac{\partial(\mathcal{F},\mathcal{G})}{\partial(y,q)}+q\frac{\partial(\mathcal{F},\mathcal{G})}{\partial(z,q)}+p_y\frac{\partial(\mathcal{F},\mathcal{G})}{\partial(p,q)}=0. \tag{5.25}$$

Finally, we add Eqs. 5.24 and 5.25, using the condition for exactness, Eq. 5.23. to get the *bilinear concomitant*

$$[\mathcal{F},\ \mathcal{G}]=\frac{\partial(\mathcal{F},\mathcal{G})}{\partial(x,p)}+\frac{\partial(\mathcal{F},\mathcal{G})}{\partial(y,q)}+p\frac{\partial(\mathcal{F},\mathcal{G})}{\partial(z,p)}+q\frac{\partial(\mathcal{F},\mathcal{G})}{\partial(z,q)}=0. \tag{5.26}$$

It is also known as the *bracket*. What we have shown is that for $\mathcal{F}$ and $\mathcal{G}$ to have a solution in common is that $[\mathcal{F},\ \mathcal{G}]=0$. This condition is sufficient as well.

5.4 Non-Linear Equation: The Method of Lagrange and Charpit

To get a general solution to the non linear, first order partial differential equation we use the method of Lagrange and Charpit[5] as extended by Jacobi.[6] Refer now to Eq. 5.18

$$\mathcal{F}(x,\ y,\ z,\ p,\ q)=0.$$

We first find a second partial differential equation

$$\mathcal{G}(x,\ y,\ z,\ p,\ q)=0,$$

that has the same solutions as $\mathcal{F}=0$. To do this we use the bilinear concomitant, Eq. 5.26 which we expand to get

$$\left(\frac{\partial\mathcal{F}}{\partial x}\frac{\partial\mathcal{G}}{\partial p}-\frac{\partial\mathcal{F}}{\partial p}\frac{\partial\mathcal{G}}{\partial x}\right)+\left(\frac{\partial\mathcal{F}}{\partial y}\frac{\partial\mathcal{G}}{\partial q}-\frac{\partial\mathcal{F}}{\partial q}\frac{\partial\mathcal{G}}{\partial y}\right) \tag{5.27}$$
$$+p\left(\frac{\partial\mathcal{F}}{\partial z}\frac{\partial\mathcal{G}}{\partial p}-\frac{\partial\mathcal{F}}{\partial p}\frac{\partial\mathcal{G}}{\partial z}\right)+q\left(\frac{\partial\mathcal{F}}{\partial z}\frac{\partial\mathcal{G}}{\partial p}-\frac{\partial\mathcal{F}}{\partial q}\frac{\partial\mathcal{G}}{\partial z}\right)=0,$$

[5]Forsyth 1959, Vol. V, pp. 156–159. Forsyth 1996, pp. 420–429. Cohen 1933, pp. 263–271.
[6]Forsyth 1996, pp. 430–439.

which we rearrange in the following way:

$$\left(\frac{\partial \mathcal{F}}{\partial x}+p\frac{\partial \mathcal{F}}{\partial z}\right)\frac{\partial \mathcal{G}}{\partial p}+\left(\frac{\partial \mathcal{F}}{\partial y}+q\frac{\partial \mathcal{F}}{\partial z}\right)\frac{\partial \mathcal{G}}{\partial q}$$
$$-\frac{\partial \mathcal{F}}{\partial p}\frac{\partial \mathcal{G}}{\partial x}-\frac{\partial \mathcal{F}}{\partial q}\frac{\partial \mathcal{G}}{\partial y}-\left(p\frac{\partial \mathcal{F}}{\partial p}+q\frac{\partial \mathcal{F}}{\partial q}\right)\frac{\partial \mathcal{G}}{\partial z}=0, \tag{5.28}$$

a linear partial differential equation for $\mathcal{G}$ and having $x,\ y,\ z,\ p$ and q as its independent variables. As always subscripts denote derivatives. From Eq. 5.6, its characteristic equations are

$$\frac{dp}{\mathcal{F}_x+p\mathcal{F}_z}=\frac{dq}{\mathcal{F}_y+q\mathcal{F}_z}=\frac{-dx}{\mathcal{F}_p}=\frac{-dy}{\mathcal{F}_q}=\frac{-dz}{p\mathcal{F}_p+q\mathcal{F}_q}. \tag{5.29}$$

Any characteristic of Eq. 5.29 is a solution of the bilinear concomitant, Eq. 5.26 and therefore can serve as the function $\mathcal{G}$ in the simultaneous pair given in Eqs. 5.18 and 5.19. These are then solved for p and q that are then plugged into the total differential equation given in Eq. 5.21, whose exactness is assured by Eq. 5.22 that was used in its derivation. The solution of this is called the *complete integral.* It depends on arbitrary constants, not on an arbitrary function and therefore cannot be a general solution.

5.5 The General Solution

The complete integral involves, in the case of equations with two independent variables, two arbitrary constants. The general solution of partial differential equations must involve exactly one arbitrary function of several variables; one less than the number of independent variables. So the complete integral is really not at all complete.

Suppose we replace the arbitrary constants by arbitrary functions of the independent variables. But a general solution of a first order equation can have no more than one arbitrary function; we may therefore apply any convenient condition to these arbitrary functions to reduce them by one.

The following example demonstrates the method. Let $\mathcal{F}$ be partial differential equation; at the same time it is a function of z, p and q

$$\mathcal{F}\equiv z-pq=0. \tag{5.30}$$

Its derivatives are

$$\mathcal{F}_x=\mathcal{F}_y=0, \qquad \mathcal{F}_z=1, \qquad \mathcal{F}_p=-q, \qquad \mathcal{F}_q=-p, \tag{5.31}$$

so that the characteristic equations become, from Eq. 5.29

$$\frac{dp}{p}=\frac{dq}{q}=\frac{dx}{q}=\frac{dz}{pq}. \tag{5.32}$$

From the first two members we get $qdp-pdq=0$ which results in

$$p=aq, \tag{5.33}$$

where a is a constant of integration. Substitute this back into Eq. 5.30 to get $q = \sqrt{z/a}$ so that the total differential equation becomes

$$dz = \sqrt{az}\, dx + \sqrt{z/a}\, dy,$$

or

$$\frac{dz}{\sqrt{z}} = \sqrt{a}\, dx + \sqrt{1/a}\, dy, \tag{5.34}$$

whose solution is

$$z = \frac{1}{4a}(ax + y + b)^2 \tag{5.35}$$

where b is a second constant of integration introduced in the quadrature of Eq. 5.34. This is the complete integral.

Suppose we instead took the second and third members of the characteristic equations, Eq. 5.32, resulting in $q = x+c$, which, when substituted back into Eq. 5.30, yields $z = p(x+c)$. The total differential equation is then

$$dz = \frac{z}{x+c} dx + (x+c)dy, \tag{5.36}$$

which rearranges itself into

$$\frac{(x+c)dz - zdx}{(x+c)^2} = dy, \tag{5.37}$$

whose solution is

$$z = (x+c)(y+d), \tag{5.38}$$

a complete integral distinctly different from that given in Eq. 5.35. The complete integral is then not unique. We will show that each leads to the same general solution.

Take the first complete integral from Eq. 5.35 and replace the two constants of integration by arbitrary functions $a(x,\, y)$ and $b(x,\, y)$; thus

$$4a(x,\, y)z = [a(x,\, y)x + y + b(x,\, y)]^2. \tag{5.39}$$

As it stands this solution contains two arbitrary functions, $a(x,\, y)$ and $b(x,\, y)$. But a first order partial differential equation must have a general solution that involves no more than one. We may then impose any convenient restrictions on these two functions. Moreover it turns out that with the two new functions Eq. 5.39 no longer satisfies the given equation. We choose constraints to correct this.

First we calculate the two derivatives of z

$$
\begin{aligned}
z_x &= \frac{1}{2a}[(ax+y+b)(a+a_x x+b_x) - 2za_x] \\
&= \frac{1}{a}\left[\sqrt{az}(a+a_x x+b_x) - za_x\right] \\
&= \sqrt{az} + \frac{1}{a}\left[\sqrt{az}(a_x x+b_x) - za_x\right] \\
&= p + \frac{1}{a}\left[a_x(x\sqrt{az}-z) + b_x\sqrt{az}\right] \\
z_y &= q + \frac{1}{a}\left[a_y(x\sqrt{az}-z) + b_y\sqrt{az}\right].
\end{aligned}
\tag{5.40}
$$

Next, we choose a and b in such a way that $z_x = p$ and $z_y = q$, thus assuring that the original differential equation, Eq. 5.30, is satisfied. That leaves the following conditions

$$
\begin{cases}
a_x(x\sqrt{az}-z) + b_x\sqrt{az} = 0 \\
a_y(x\sqrt{az}-z) + b_y\sqrt{az} = 0,
\end{cases}
\tag{5.41}
$$

which, when cast in matrix form, becomes

$$
\begin{pmatrix} a_x & b_x \\ a_y & b_y \end{pmatrix}
\begin{pmatrix} x\sqrt{az}-z \\ \sqrt{az} \end{pmatrix}
=
\begin{pmatrix} 0 \\ 0 \end{pmatrix}.
\tag{5.42}
$$

This is a linear system with two possible interpretations.[7] If the determinant of coefficients does not vanish then there can be only a trivial solution, namely, $z = 0$, the *singular solution*. Now suppose the determinant of coefficients is equal to zero. This determinant is a Jacobian; its vanishing implies the existence of a functional relationship between a and b. Suppose that relationship is $b = f(a)$ so that $b_x = a_x f_a$ and $b_y = a_y f_a$. Plugging these into the two equations of Eq. 5.41 results in the single equation

$$
x\sqrt{az} - z + f'\sqrt{az} = 0.
\tag{5.43}
$$

Now we adjoin Eqs. 5.35 to 5.43 to get the *general solution*

$$
\begin{cases}
z = \dfrac{1}{2a}[ax+y+f(a)]^2 \\
x + f' = \sqrt{z/a}.
\end{cases}
\tag{5.44}
$$

Unlike the linear equation, this general solution contains not only the arbitrary function $f(a)$ but its first derivative as well. It also consists of two separate equations. To obtain a specific solution from Eq. 5.44, as in the case of an initial value or a boundary value, one needs only to find an appropriate function f and then eliminate a. By choosing a different f we obtain

[7] Dickson 1939, pp. 131–132.

a different particular solution, *et cetera, ad infinitum*. One can think of the general solution as consisting of the totality of these particular solutions. There is yet another approach to the idea of a general solution that we will use later.

Now return to the other complete integral given in Eq. 5.39, replace c and d by arbitrary functions, and take its two partial derivatives

$$\begin{aligned} z_x &= (y+d) + c_x(y+d) + d_x(x+c) \\ &= p + c_x(y+d) + d_x(x+c) \\ z_y &= (x+c) + c_y(y+d) + d_y(x+c) \\ &= q + c_y(y+d) + d_y(x+c). \end{aligned} \tag{5.45}$$

Again, we choose c and d so that $z_x = p$ and $z_y = q$ to get

$$\begin{cases} c_x(y+d) + d_x(x+c) = 0 \\ c_y(y+d) + d_y(x+c) = 0. \end{cases} \tag{5.46}$$

For the general solution, the determinant of coefficients, the Jacobian, must vanish, a relationship then must exist between c and d, say $d = g(c)$, which yields, from Eq. 5.46, $y + g + (x+c)g' = 0$. This and Eq. 5.38 constitute a different form of the general solution

$$\begin{cases} z = (x+c)[y+g(c)] \\ \quad y + g(c) + (x+c)g'(c) = 0. \end{cases} \tag{5.47}$$

Note that from these two relations a third can be obtained

$$z = -g'(x+c)^2. \tag{5.48}$$

The point here is that these two forms of the general solution, given in Eq. 5.44 and in Eqs. 5.47 and 5–48 are really the same. Here if we substitute these expressions, $c = (y+f)/a$, $g' = -a/4$ and $g = (ax - y + f)/2$ into the Eqs. 5.47 and 5.48 we will get the first equation of Eq. 5.45; the second will be satisfied. The point of all this is to demonstrate that the general solution can be expressed in a number of equivalent forms and that one can be transformed into the other.

5.6 The Extension to Three Independent Variables

Now suppose our first order equation possesses three independent variables, x, y, and z; thus

$$\mathcal{F}(x,\ y,\ z,\ \phi,\ \phi_x,\ \phi_y,\ \phi_z) = 0. \tag{5.49}$$

Again we replace the three partial derivatives by three symbols

$$p_1 = \phi_x, \qquad p_2 = \phi_y, \qquad p_3 = \phi_z,$$

so that the differential equation in Eq. 5.49 takes the form of an algebraic equation in seven variables

$$\mathcal{F}(x,\ y,\ z,\ \phi,\ p_1,\ p_2,\ p_3) = 0. \tag{5.50}$$

The characteristic equations now take the form

$$\begin{aligned} &\frac{dp_1}{\mathcal{F}_x + p_1\mathcal{F}_\phi} = \frac{dp_2}{\mathcal{F}_y + p_2\mathcal{F}_\phi} = \frac{dp_2}{\mathcal{F}_z + p_3\mathcal{F}_\phi} \\ &= \frac{-dx}{\mathcal{F}_{p1}} = \frac{-dy}{\mathcal{F}_{p2}} = \frac{-dz}{\mathcal{F}_{p3}} = \frac{-d\phi}{p_1\mathcal{F}_{p_1} + p_2\mathcal{F}_{p_2} + p_3\mathcal{F}_{p_3}}. \end{aligned} \tag{5.51}$$

From these we extract two independent characteristics, say, $\mathcal{G}_1$ and $\mathcal{G}_2$, which, together with $\mathcal{F}$ give us three algebraic equations

$$\begin{cases} \mathcal{F}(x,\ y,\ z,\ \phi,\ p_1,\ p_2,\ p_3) = 0 \\ \mathcal{G}_1(x,\ y,\ z,\ \phi,\ p_1,\ p_2,\ p_3) = 0 \\ \mathcal{G}_2(x,\ y,\ z,\ \phi,\ p_1,\ p_2,\ p_3) = 0, \end{cases} \tag{5.52}$$

which we solve for $p_1,\ p_2$ and p_3

$$\begin{cases} p_1 = A_1(x,\ y,\ z,\ \phi) \\ p_2 = A_2(x,\ y,\ z,\ \phi) \\ p_3 = A_3(x,\ y,\ z,\ \phi), \end{cases} \tag{5.53}$$

which is substituted into the total differential equation

$$d\phi = p_1 dx + p_2 dy + p_3 dz = A_1 dx + A_2 dy + A_3 dz, \tag{5.54}$$

whose solution is the complete integral. The general solution is obtained from the complete integral in the same way as given in the examples.

5.7 The Eikonal Equation. The Complete Integral

We use the method of Lagrange and Charpit to solve the eikonal equation, given in vector form in Eq. 3.65 and in scalar form in Eq. 5.1.

We restrict ourselves to homogeneous, isotropic media, otherwise the rather simple characteristic equation that follows will be complicated by the presence of the derivatives of n, the refractive index of the medium.

Moreover, the function $\phi(x,\ y,\ z)$ is equal to the optical path length from some arbitrary but fixed object point, through an optical system, to the point $(x,\ y,\ z)$. By setting this equal to a constant the result is the equation of a surface with the property that each point is equidistant, in terms of optical path length, from that object point. This surface is then a *surface of constant phase* or, more simply, a *wavefront*.

In what follows we use the definitions introduced in Eq. 5.51 so that the eikonal equation, Eq. 5.1, takes the form of an algebraic equation

$$\mathcal{F} \equiv p_1^2 + p_2^2 + p_3^2 - n^2 = 0. \tag{5.55}$$

Then

$$\mathcal{F}_{p_1} = 2p_1,\ \mathcal{F}_{p_2} = 2\,p_2,\ \mathcal{F}_{p_3} = 2\,p_3,\ \mathcal{F}_x = \mathcal{F}_y = \mathcal{F}_z = \mathcal{F}_\phi = 0, \tag{5.56}$$

so that the characteristic equations become

$$\frac{dp_1}{0} = \frac{dp_2}{0} = \frac{dp_3}{0} == \frac{-dx}{2p_1} = \frac{-dy}{2p_2} = \frac{-dz}{2p_3} = \frac{-d\phi}{2(p_1^2 + p_2^2 + p_3^2)}. \tag{5.57}$$

We choose the simplest two, $dp_2 = 0$, and $dp_3 = 0$; these lead to

$$p_2 = v, \qquad p_3 = w, \tag{5.58}$$

where v and w are constants of integration. These are substituted into the original equation, Eq. 5.56 which yields

$$p_1 = \sqrt{n^2 - v^2 - w^2}, \tag{5.59}$$

so that

$$d\phi = \sqrt{n^2 - v^2 - w^2}dx + vdy + wdz. \tag{5.60}$$

Integrating this yields the complete integral

$$\phi = x\sqrt{n^2 - v^2 - w^2} + yv + zw + k, \tag{5.61}$$

where k is a constant of integration. Luneburg[8] came this far but unfortunately stopped.

As we will see later the general solution of the eikonal equation is unable to represent plane waves but the complete integral can. Luneburg did develop a theory of optical wavefronts, based only on plane waves, as singularities propagating through a medium.

Now ϕ is a function equal to the optical path length of a ray from some arbitrary starting point to $(x,\ y,\ z)$. By setting ϕ equal to a constant we get the equation of the locus of points with constant optical path length that start from an arbitrary but fixed object point, that is to say, a *wavefront*. If we designate this constant by ns, where s is a geometrical distance, we obtain the equation for a one-parameter family of wavefronts

$$\phi = ns, \tag{5.62}$$

where s is the distance parameter.

[8] Luneburg 1966, pp. 21–57.

5.8 The Eikonal Equation. The General Solution

Now we pass to the general solution. Assume that v, w and k are functions of x, y and z. Then the derivatives of Eq. 5.62 are

$$\begin{cases} \phi_x = \sqrt{n^2 - v^2 - w^2} - x\dfrac{vv_x + ww_x}{\sqrt{n^2 - v^2 - w^2}} + yv_x + zw_x + k_x, \\ \phi_y = \qquad v \qquad - x\dfrac{vv_y + ww_y}{\sqrt{n^2 - v^2 - w^2}} + yv_y + zw_y + k_y, \\ \phi_z = \qquad w \qquad - x\dfrac{vv_z + ww_z}{\sqrt{n^2 - v^2 - w^2}} + yv_z + zw_z + k_z. \end{cases} \tag{5.63}$$

For $\mathcal{F}_x = p_1$, etc. it must be that

$$\begin{cases} (y - vx/u)v_x + (z - wx/u)w_x + k_x = 0, \\ (y - vx/u)v_y + (z - wx/u)w_y + k_y = 0, \\ (y - vx/u)v_z + (z - wx/u)w_z + k_z = 0, \end{cases} \tag{5.64}$$

where we have taken $u = \sqrt{n^2 - v^2 - w^2}$.

Write Eq. 5.64 in matrix form

$$\begin{pmatrix} v_x & w_x & k_x \\ v_y & w_y & k_y \\ v_z & w_z & k_z \end{pmatrix} \begin{pmatrix} y - vx/u \\ z - wx/u \\ 1 \end{pmatrix} = \begin{pmatrix} 0 \\ 0 \\ 0 \end{pmatrix}. \tag{5.65}$$

A non trivial solution will bc obtained if and only if the determinant of coefficients is zero. That determinant is a Jacobian; if the Jacobian vanishes then there exists a relation between the functions v, w and k. Suppose that relationship is given by $k = k(v,\ w)$. Inserting this into Eq. 5.64 yields

$$\begin{cases} (y - vx/u + k_v)v_x + (z - wx/u + k_w)w_x = 0, \\ (y - vx/u + k_v)v_y + (z - wx/u + k_w)w_y = 0, \\ (y - vx/u + k_v)v_z + (z - wx/u + k_w)w_z = 0, \end{cases} \tag{5.66}$$

which, in turn, results in

$$\begin{cases} y - \ vx/u + k_v \ = 0, \\ z - wx/u + k_w = 0. \end{cases} \tag{5.67}$$

Equation 5.67, when conjoined with the complete integral, Eq. 5.63, comprises the general solution

$$\begin{cases} \phi = x\sqrt{n^2 - v^2 - w^2} + yv + zw + k(v,\ w) = ns, \\ y - vx/u + k_v = 0, \\ z \ - wx/u + k_w \ - 0. \end{cases} \tag{5.68}$$

To apply this, choose some specific function of two variables for k and substitute it into Eq. 5.68. Then, by eliminating v and w (and therefore u) between the three equations the result is, say, $z = z(x,\ y,\ s)$, the equation of a train of wavefronts.

However we will use an alternative technique. Recall that the eikonal ϕ is the optical path length from some arbitrary but fixed object point to the point $(x,\ y,\ z)$ and that by setting ϕ equal to a constant we get the equation of a surface whose points all have the same optical path length from the object point. In particular we set $\phi(x,\ y,\ z) = n\,s$ as we already have in Eq. 5.68. We retain the parameters v and w and solve Eq. 5.70 for x, y and z as functions of v and w. The solution of the simultaneous system in Eq. 5.68 is

$$\begin{cases} x = \dfrac{1}{n^2}[(ns-k)+(vk_v+wk_w)]u, \\ y = \dfrac{1}{n^2}[(ns-k)+(vk_v+wk_w)]v - k_v, \\ z = \dfrac{1}{n^2}[(ns-k)+(vk_v+wk_w)]w - k_w. \end{cases} \tag{5.69}$$

Now define these vector quantities

$$\begin{cases} \mathbf{W} = (x,\ y,\ z) \\ \mathbf{S}\ = (u,\ v,\ w) \\ \mathbf{K}\ = (0,\ k_v,\ k_w), \end{cases} \tag{5.70}$$

and the scalar quantity

$$q = (ns-k)+(vk_v+wk_w) = (ns-k)+\mathbf{S}\cdot\mathbf{K}, \tag{5.71}$$

The expression for the general wavefront can then be written in the form of a vector equation

$$\mathbf{W}(v,\ w,\ s) = \frac{q}{n^2}\mathbf{S} - \mathbf{K}. \tag{5.72}$$

The function $k(v,\ w)$, which we refer to as the *k-function*, is the arbitrary function that arises in obtaining the general solution of the eikonal equation and v and w are the parameters that appeared in the complete integral.

For each value of s, $\mathbf{W}$ in Eq. 5.72 defines a surface in space, all points of which are equidistant (measured as an optical path length) from some arbitrary but fixed initial point and is therefore a wavefront. By allowing s to vary we generate a one-parameter family of surfaces that constitutes a wavefront train.

In the k-function resides all of the geometric properties of every wavefront in the train. If the wavefront train originates at some object point, then passes through some optical system, the k-function contains all the geometrical aberrations of that lens associated with that object point.

We make the distinction here between the geometric properties of the wavefronts and other properties that arise from the physical nature of electromagnetic radiation. In Chapter 10 we will investigate this difference further.

5.9 The Eikonal Equation. Proof of the Pudding

It must be that $\mathbf{W}(v,\ w,\ s)$ given in Eq. 5.72 must satisfy $\phi(\mathbf{W}) = ns$. By taking the derivative of ϕ with respect to $v,\ w$ and s results in

$$\begin{cases} \nabla\phi \cdot \mathbf{W}_v = 0 \\ \nabla\phi \cdot \mathbf{W}_w = 0 \\ \nabla\phi \cdot \mathbf{W}_s = n. \end{cases} \tag{5.73}$$

From the first two of these equations we can get

$$(\nabla\phi \cdot \mathbf{W}_w)\mathbf{W}_v - (\nabla\phi \cdot \mathbf{W}_v)\mathbf{W}_w = \nabla\phi \times (\mathbf{W}_v \times \mathbf{W}_w) = D\nabla\phi \times \mathbf{W}_s = 0, \tag{5.74}$$

which tells us that $\nabla\phi$ and $\mathbf{W}_s$ are parallel; for some μ, $\nabla\phi = \mu\mathbf{W}_s$. From the third equation of Eq. 5.73 we can see that $\mu = n$ which leads to

$$\nabla\phi = n\mathbf{W}_s. \tag{5.75}$$

By squaring this we get back the eikonal equation.

The k-function

The function $k(u,\ v)$ is the arbitrary function that arises in the general solution of the eikonal equation and v and w are the parameters that came from the complete integral.

For each s, we have a vector function of two parameters which describes a surface in space all points of which are equidistant (measured as an optical path length) from some arbitrary but fixed starting point and is therefore a wavefront. By allowing s to vary we get a one-parameter family of surfaces that constitute a wavefront train.

In the k-function resides all of the geometric properties of every wavefront in the train. If the wavefront train is generated by light from some object point that passes through a lens then its monochromatic geometrical aberrations are all contained in the k-function. We make the distinction here between the geometric properties of the wavefronts and the diffraction effects that arise from the physical nature of electromagnetic radiation. In Chapter 10 we will investigate this difference further.

Each of these wavefronts corresponds to a fixed value of s and is therefore represented by a vector of two parameters. This permits us to analyze the wavefront from the point of view of the differential geometry of surfaces that we encountered in Chapter 4. This is the topic of the next chapter, Chapter 6.

Part II

The k-function

The Mathematics of Geometrical and Physical Optics: The k-function and its Ramifications. O.N. Stavroudis

ISBN: 3-527-40448-1

6 The Geometry of Wave Fronts

In Chapter 4 we found that surfaces can be represented by vector functions of two parameters and that the choice of parameters had no effect on their geometric structure. For example, the unit normal vector $\mathbf{N}$, because it is a surface property, is invariant with respect to parameter transformations. In addition we found that at each regular point of a smooth surface the curvature of a normal section assumed extremum values in two orthogonal directions. These curvatures are called the *principal curvatures* at that point and the two directions are referred to as *the principal directions*.

In Chapter 5 we obtained a general solution of the eikonal equation for homogeneous, isotropic media that we interpreted as one-parameter families of wavefronts. These depend on a length parameter s that is associated with individual members of the family, and v and w, two of the reduced ray direction cosines. The general solution of the eikonal equation includes an arbitrary function, the k-function, that depends on v and w but not on s. The geometric properties of each wavefront in the wavefront train depends entirely on the k-function and its first derivatives.

We now have an analytic expression in vector form for wavefront trains in homogeneous, isotropic media. In this chapter we apply the differential geometry of surfaces to the general expression for a wavefront. This leads to the calculation of the wavefronts' principal curvatures and principal directions. This is the principal objective of this chapter.

6.1 Preliminary Calculations

Now to get to specifics. From Chapter 5, we have the solution of the eikonal equation

$$\text{(5.72)} \qquad \mathbf{W}(v,\, w;\, s) = \frac{q}{n^2}\mathbf{S} - \mathbf{K},$$

where

$$\text{(5.71)} \qquad q = (ns-k)+(vk_v+wk_w) = (ns-k)+\mathbf{S}\cdot\mathbf{K},$$

and where

$$\text{(5.70)} \qquad \begin{cases} \mathbf{S} = (u,\, v,\, w) & \mathbf{S}^2 = n^2 \\ \mathbf{K} = (0,\, k_v,\, k_w). \end{cases}$$

As always, subscripts signal ordinary or partial differentiation.

The Mathematics of Geometrical and Physical Optics: The k-function and its Ramifications. O.N. Stavroudis

ISBN: 3-527-40448-1

We will need the three partial derivatives of $\mathbf{W}$. These are

$$\begin{cases} \mathbf{W}_v = \dfrac{1}{n^2}(\mathbf{S}\cdot\mathbf{K}_v)\mathbf{S} - \mathbf{K}_v + \dfrac{q}{n^2}\,\mathbf{S}_v \\ \mathbf{W}_w = \dfrac{1}{n^2}(\mathbf{S}\cdot\mathbf{K}_w)\mathbf{S} - \mathbf{K}_w + \dfrac{q}{n^2}\,\mathbf{S}_w \\ \mathbf{W}_s = \frac{1}{n}\mathbf{S}, \end{cases} \tag{6.1}$$

where q is defined in Eq. 5.71. Recall from Eq. 5.70 that $\mathbf{S} = (u,\ v,\ w)$ where we have taken $u = \sqrt{n^2 - v^2 - w^2}$. The derivatives of $\mathbf{S}$ are then

$$\begin{cases} \mathbf{S}_v = -\dfrac{1}{u}(v, -u,\ 0) = -\dfrac{1}{u}(\mathbf{S}\times\mathbf{Z}) \\ \mathbf{S}_w = -\dfrac{1}{u}(w,\ 0, -u) = \ \dfrac{1}{u}(\mathbf{S}\times\mathbf{Y}). \end{cases} \tag{6.2}$$

The vectors $\mathbf{X}$, $\mathbf{Y}$ and $\mathbf{Z}$ are the three unit vectors in the direction of the three coordinate axes.

On substituting Eq. 6.2 into Eq. 6.1 we get

$$\begin{cases} \mathbf{W}_v = \dfrac{1}{n^2}\mathbf{S}\times\left[(\mathbf{S}\times\mathbf{K}_v) - \dfrac{q}{u}\mathbf{Z}\right] \\ \mathbf{W}_w = \dfrac{1}{n^2}\mathbf{S}\times\left[(\mathbf{S}\times\mathbf{K}_w) + \dfrac{q}{u}\mathbf{Y}\right]. \end{cases} \tag{6.3}$$

The first fundamental quantities that were defined in Eq. 4.4, in this context, become

$$E = \mathbf{W}_v^2 \qquad F = \mathbf{W}_v\cdot\mathbf{W}_w \qquad G = \mathbf{W}_w^2. \tag{6.4}$$

We use Eq. 6.3 to calculate E.

$$\begin{aligned} n^4u^2E &= \{\mathbf{S}\times[u(\mathbf{S}\times\mathbf{K}_v) - q\mathbf{Z}]\}^2 \\ &= n^2[u(\mathbf{S}\times\mathbf{K}_v) - q\mathbf{Z}]^2 - \{\mathbf{S}\cdot[u(\mathbf{S}\times\mathbf{K}_v) - q\mathbf{Z}]\}^2 \\ &= n^2\{n^2u^2(k_{vv}^2 + k_{vw}^2) - (vk_{vv} + wk_{vw})^2 \\ &\qquad -2u^2qk_{vv} + q^2\} - q^2w^2 \\ &= n^2u^2[k_{vv}H - (n^2 - w^2)T^2 - 2qk_{vv}] + q^2(n^2 - w^2) \end{aligned} \tag{6.5}$$

where

$$H = (n^2 - v^2)k_{vv} + (n^2 - w^2)k_{ww} - 2vwk_{vw}, \tag{6.6}$$

and

$$T^2 = k_{vv}k_{ww} - k_{vw}^2. \tag{6.7}$$

We also calculate F and G in the same way to get

$$\begin{cases} n^4u^2E = (q^2 - n^2u^2T^2)(n^2 - w^2) + n^2u^2(H - 2q)k_{vv} \\ n^4u^2F = (q^2 - n^2u^2T^2)vw \qquad\qquad + n^2u^2(H - 2q)k_{vw} \\ n^4u^2G = (q^2 - n^2u^2T^2)(n^2 - v^2) + n^2u^2(H - 2q)k_{ww}. \end{cases} \tag{6.8}$$

Also from Chapter 4 we have the expression for the element of surface area,

$$D^2 = (\mathbf{W}_v \times \mathbf{W}_w)^2 = EG - F^2, \tag{4.8}$$

and the unit normal vector,

$$\mathbf{N} = (\mathbf{W}_v \times \mathbf{W}_w)/D. \tag{4.7}$$

To calculate D we use the derivatives in Eq. 6.3, the first fundamental forms in Eq. 6.5 and the quantities defined in Eqs. 6.6 and 6.7 in the following:

$$\begin{aligned} &n^8u^4(EG - F^2) \\ &= [(q^2 - n^2u^2T^2)(n^2 - w^2) + n^2u^2(H - 2q)k_{vv}] \\ &\quad \times [(q^2 - n^2u^2T^2)(n^2 - v^2) + n^2u^2(H - 2q)k_{ww}] \\ &\quad - [(q^2 - n^2u^2T^2)vw + n^2u^2(H - 2q)k_{vw}]^2 \\ &= n^2u^2[(q^2 - n^2u^2T^2)^2 \\ &\qquad + (q^2 - n^2u^2T^2)(H - 2q)H + n^2u^2(H - 2q)^2T^2]. \\ &= n^2u^2[q^2(q - H)^2 + 2n^2u^2T^2q(q - H) + n^4u^4T^4] \\ &= n^2u^2[q(q - H) + n^2u^2T^2]^2, \end{aligned} \tag{6.9}$$

a perfect square. It follows that

$$D = \frac{1}{n^3u}(q^2 - Hq + n^2u^2T^2). \tag{6.10}$$

Now we come to the second fundamental quantities defined as,

$$\begin{cases} L \;= \mathbf{N} \cdot \mathbf{W}_{vv} = -\mathbf{N}_v \cdot \mathbf{W}_v \\ M = \mathbf{N} \cdot \mathbf{W}_{vw} = -\mathbf{N}_v \cdot \mathbf{W}_w = -\mathbf{N}_w \cdot \mathbf{W}_v \\ N \;= \mathbf{N} \cdot \mathbf{W}_{ww} = -\mathbf{N}_w \cdot \mathbf{W}_w. \end{cases} \tag{4.31}$$

Now we make use of Eq. 6.1 in which $\mathbf{W}_v$ and $\mathbf{W}_w$ are determined. Since $\mathbf{N} = \mathbf{S}/n$ we can use Eq. 6.2 to get $\mathbf{N}_v$ and $\mathbf{N}_w$ which leads us to the second fundamental quantities

$$\begin{cases} n^3u^2L \;= n^2u^2k_{vv} - q(n^2 - w^2) \\ n^3u^2M = n^2u^2k_{vw} - qvw \\ n^3u^2N = n^2u^2k_{ww} - q(n^2 - v^2), \end{cases} \tag{6.11}$$

in which we use Eq. 4.31.

The direction vectors are,

$$\mathbf{T}_\pm = \frac{\mathbf{W}_v + \lambda_\pm \mathbf{W}_w}{\sqrt{E + 2F\lambda_\pm + G\lambda_\pm^2}}. \tag{4.14}$$

If the values of λ are the two roots of the quadratic polynomial,

$$(EM - FL) + (EN - GL)\lambda + (FN - GM)\lambda^2 = 0 \tag{4.19}$$

then the directions given in Eq. 4.14 are principal directions. In that case the two equivalent versions of the formula for curvature are,

$$\frac{1}{\rho_\pm} = \frac{L + M\lambda_\pm}{E + F\lambda_\pm} = \frac{M + N\lambda_\pm}{F + G\lambda_\pm}. \tag{4.25}$$

From Eqs. 6.5 and 6.11 we get the coefficients of the polynomial in Eq. 4.19

$$\begin{cases} n^2u(FN - GM) = D\left[(n^2 - v^2)k_{vw} - vwk_{ww}\right] \\ n^2u(EN - GL) \;= D\left[(n^2 - v^2)k_{vv} - (n^2 - w^2)k_{ww}\right] \\ n^2u(EM - FL) \;= D\left[-(n^2 - w^2)k_{vw} + vwk_{vv}\right], \end{cases} \tag{6.12}$$

so that Eq. 4.19 takes the form

$$\begin{aligned} \left[(n^2 - v^2)k_{vw} - vwk_{ww}\right]\lambda^2 + \left[(n^2 - v^2)k_{vv} - (n^2 - w^2)k_{ww}\right]\lambda \\ - (n^2 - w^2)k_{vw} + vwk_{vv} = 0, \end{aligned} \tag{6.13}$$

whose solution is

$$\lambda_\pm = \frac{-\left[(n^2 - v^2)k_{vv} - (n^2 - w^2)k_{ww}\right] \pm \mathcal{S}}{2\left[(n^2 - v^2)k_{vw} - vwk_{ww}\right]}, \tag{6.14}$$

where

$$\mathcal{S}^2 = H^2 - 4n^2u^2T^2. \tag{6.15}$$

The quantities H and T^2 are defined in Eqs. 6.6 and 6.7.

Now refer back to Eq. 4.25, a pair of equations for the two principal radii of curvature, in terms of the parameter λ. We eliminate λ between these to get a quadratic polynomial in ρ

$$(LN - M^2)\rho^2 + \rho(2FM - EN - GL) + (EG - F^2) = 0, \tag{6.16}$$

the coefficients of which turn out to be, using Eqs. 6.5 and 6.11

$$\begin{cases} LN - M^2 \qquad\qquad = D/nu \\ EN + GL - 2FM = D(H - 2q)/n^2u \\ EG - F^2 \qquad\qquad = D^2. \end{cases} \tag{6.17}$$

On substituting these relations into the polynomial in Eq. 6.16 it becomes

$$n\rho^2 - (H - 2q)\rho + n^2 uD = 0, \tag{6.18}$$

whose solution is

$$\rho_{\pm} = \frac{1}{2n}[H - 2q \pm \mathcal{S}] = -\frac{1}{n}\left(q - \frac{H \pm \mathcal{S}}{2}\right), \tag{6.19}$$

where $\mathcal{S}$ is given in Eq. 6.15. Note that $\mathcal{S}$ is the discriminant of both Eq. 6.13 and Eq. 6.18. The two values of ρ so obtained are the two principal radii of curvature.

From Eq. 6.19 we can see that the expression for D in Eq. 6.10, factors into

$$D = \frac{1}{n^3 u}\left(q - \frac{H + \mathcal{S}}{2}\right)\left(q - \frac{H - \mathcal{S}}{2}\right) = \frac{1}{nu}\rho_+\rho_-, \tag{6.20}$$

which shows that D vanishes whenever one of the principal radii of curvatures vanishes.

6.2 The Caustic Surface

Among the artifacts of geometrical optics, rays possess no physical reality; wavefronts can not be seen, their presence must be observed indirectly by means of interferometry. On the other hand the caustic surface is indeed real and can be seen and photographed.[1] Yet of the *materia* of geometrical optics it is the least understood and almost completely ignored.

The caustic can be defined as the envelope of the system of the orthotomic rays encountered in Chapter 3. The definition that we will use here is that the caustic is the locus of the principal centers of curvature of a wavefront. This reveals most clearly the arcane properties of the caustic and imply its possible use in optical design.[2]

Select one wavefront in the train and on that wavefront select one point; in doing so we fix v, w, and s. For these values there will be two values of ρ from Eq. 6.19, the two principal radii of curvature and therefore distances to the two principle centers of curvature. The point on the wavefront is given in the usual way from Eqs. 5.71 and 5.72

$$\mathbf{W}(v,\ w,\ s) = \frac{1}{n^2}\big[ns - k + (\mathbf{S}\cdot\mathbf{K})\big]\mathbf{S} - \mathbf{K}.$$

Through that point passes exactly one ray. We move along that ray the two distances ρ_+ and ρ_-. This takes us to the two principal centers of curvature. Perhaps it would be clearer to say that if $\mathbf{C}_{\pm}$ represents the two principal center of curvature then $\mathbf{C}_{\pm} = \mathbf{W} + (\rho_{\pm}/n)\mathbf{S}$. By making the obvious substitution using Eq. 6.19 we get

$$\mathbf{C}_{\pm} = \frac{1}{2n^2}(H \pm \mathcal{S})\mathbf{S} - \mathbf{K}. \tag{6.21}$$

Now we allow v and w to vary and generate the locus of the principal centers of curvature or the caustic surface. We can see from all this that the caustic is a surface of two sheets each

[1] Cagnet et al. 1962

[2] Stavroudis 1996.

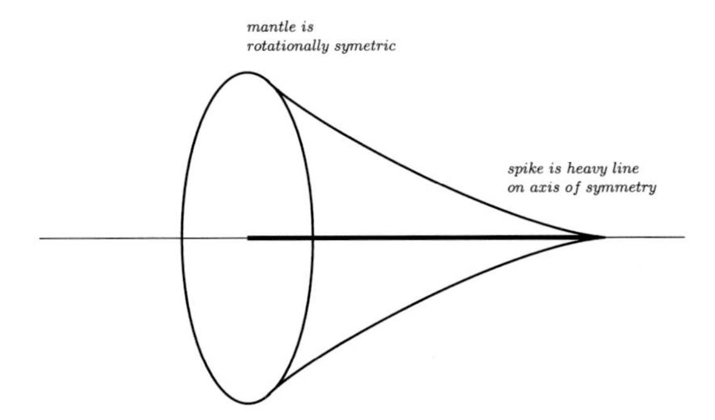

Figure 6.1: *The mantle and the spike.*

corresponding to one of the branches of $\mathbf{C}_{\pm}$. Moreover $\mathbf{C}_{\pm}(v,\ w)$ is independent of s and therefore is a property of the entire wavefront train. The shape of the caustic depends only on the k-function and its first and second derivatives.

To visualize this consider the image of an axial object point produced by a rotationally symmetric optical system. In this case $\mathbf{C}_{-}$ degenerates into the segment of a straight line lying on the axis of symmetry. This we refer to as the *spike*. Surrounding the spike is the *mantle*, a rotationally symmetric surface that is centered on the axis. This is shown in Fig. 6.1.

We define K, the *Gaussian curvature* at a point, as the product of the two principal curvatures at that point. Those for a wavefront are, $1/\rho_{+}$ and $1/\rho_{-}$ so that $K = 1/\rho_{+}\rho_{-}$. From Eqs 6.19 and 6.20

$$K = \frac{1}{nuD} = \frac{n^2}{q^2 - qH + n^2u^2T^2}, \tag{6.22}$$

where q, H and T^2 are defined in Eqs. 5.71, 6.6 and 6.7, respectively.

Now refer back to Eq. 6.20 in which D, the wavefront element of area, and Eq. 6.22, which relates D and K, the Gaussian curvature. Where a wavefront passes through a principal center of curvature the quantity ρ vanishes. It follows that D must also vanish and the Gaussian curvature becomes infinite. We can interpret this as a crease or fold of the wavefront that forms a cusp. This means that we can add to our definitions of caustic that it is the *cusp locus* of the wavefront train.

6.3 Special Surfaces I: Plane and Spherical Wavefronts

There are two special cases that are nevertheless common. In the case of the plane wavefront we are faced with the same paradox that led Hamilton to distinguish several different classes of eikonals, what he called *characteristic functions*. Consider the point eikonal which depends on the coordinates of points in space, and the angle eikonal, that has direction cosines as arguments. The expression for wavefronts that we are using here depends on representations of the direction its associated rays. In the case of the plane wavefront, on the other hand, the ray direction is fixed and the complete integral, as in Eq. 5.61, seems the more appropriate

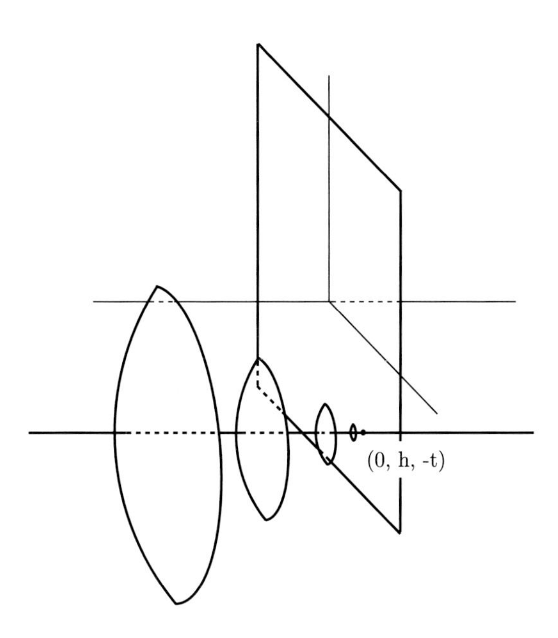

Figure 6.2: *The spherical wavefront.*

means for its representation. Indeed Luneburg[3] did this with considerable success from which he developed a theory of wave propagation distinctly different from that presented here.

The spherical wavefront is an entirely different matter in that it is completely consistent with the general solution of the eikonal equation. Figure 6.2 illustrates this. Let the point $(0,\ -h,\ -t)$ be the center of a concentric train of spherical wavefronts. Then, when $s = 0$ in Eq. 5.72, the expression for $\mathbf{W}(v,\ w,\ 0)$ must degenerate into $(0,\ -h,\ -t)$ so that

$$\mathbf{W}(v,\ w,\ 0) = \frac{1}{n^2}(-k + \mathbf{S}\cdot\mathbf{K})\mathbf{S} - \mathbf{K} = (0,\ -h,\ -t). \tag{6.23}$$

For this to happen it must be that

$$k = \mathbf{S}\cdot\mathbf{K}, \tag{6.24}$$

which indicates that in this case the k-function is homogeneous as indicated in Eq. 5.13. From Eq. 6.24 it follows that

$$k_v = h, \qquad k_w = t. \tag{6.25}$$

From these two results we get an expression for the k-function

$$k = v\,h + w\,t. \tag{6.26}$$

The expression for the train of spherical wavefronts then becomes

$$\mathbf{W}(v,\ w,\ s) = \frac{s}{n}\mathbf{S} - (0,\ h,\ t). \tag{6.27}$$

[3]Luneburg 1964, Chapter 1.

To get the derivatives of $\mathbf{W}$ we must use Eq. 6.2 which yields

$$\mathbf{W}_v = -\frac{s}{n\,u}(\mathbf{S} \times \mathbf{Z}), \qquad \mathbf{W}_w = \frac{s}{n\,u}(\mathbf{S} \times \mathbf{Y}). \tag{6.28}$$

To get D we first calculate

$$\begin{aligned}
\mathbf{W}_v \times \mathbf{W}_w &= -\frac{s^2}{n^2u^2}(\mathbf{S} \times \mathbf{Z}) \times (\mathbf{S} \times \mathbf{Y}) \\
&= -\frac{s^2}{n^2u^2}[\mathbf{S} \cdot (\mathbf{Z} \times \mathbf{Y})]\mathbf{S} \\
&= \frac{s^2}{n^2u^2}(\mathbf{S} \cdot \mathbf{X})\mathbf{S} \\
&= \frac{s^2}{n^2u}\mathbf{S}.
\end{aligned} \tag{6.29}$$

Then

$$D^2 = (\mathbf{W}_v \times \mathbf{W}_w)^2 = \frac{s^4}{n^2u^2},$$

so that

$$D = \frac{s^2}{n\,u}. \tag{6.30}$$

We will return to the spherical wavefront in Chapter 9.

6.4 Parameter Transformations

As we have stressed repeatedly in previous chapters the transformation of the surface parameters does not in any way effect the geometry of a surface;

We introduce two new parameters, p and q, into the vector function for the wavefront $\mathbf{W}$ just as we had done in Chapter 4

$$\mathbf{P}(p,\,q) = \mathbf{P}\big(v(p,\,q),\,w(p,\,q)\big). \tag{4.42}$$

By substituting $\mathbf{W}$ for $\mathbf{P}$ for the derivatives of Eq. 4.43 we get

$$\begin{cases} \mathbf{W}_p = \mathbf{W}_v\,v_p + \mathbf{W}_w\,w_p \\ \mathbf{W}_q = \mathbf{W}_v\,v_q + \mathbf{W}_w\,w_q. \end{cases} \tag{6.31}$$

In what follows lower case letters will indicate the first and second fundamental quantities associated with the parameters p and q. The first fundamental quantities are calculated from Chapter 4

$$\begin{cases} e = E\,v_p^2 + 2F\,v_p\,w_p + G\,w_p^2 \\ f = E\,v_p\,v_q + F\,(v_p\,w_q + v_q\,w_p) + G\,w_p\,w_q \\ g = E\,v_q^2 + 2F\,v_q\,w_q + G\,w_q^2. \end{cases} \tag{4.45}$$

In exactly the same way we calculate the second fundamental quantities. Again referring back to Chapter 4, in particular to,

$$(4.52)\qquad \begin{cases} l &= L\,v_p^2 + 2M\,v_p\,w_p + N w_p^2 \\ m &= L\,v_p\,v_q + M(v_p\,w_q + v_q\,w_p) + N\,w_p\,w_q \\ n &= L v_q^2 + 2M\,v_q\,w_q + N\,w_q^2. \end{cases}$$

In terms of the fundamental quantities for the new parameters, p and q, the principal curvatures are

$$\frac{1}{\rho_\pm} = \frac{l + \lambda_\pm m}{e + \lambda_\pm f} = \frac{m + \lambda_\pm n}{f + \lambda_\pm g}, \tag{6.32}$$

this from Eqs. 4.18 or 4.25. Here λ is a solution of the quadratic polynomial

$$(em - fl) + (en - gl)\lambda + (fn - gm)\lambda^2 = 0, \tag{6.33}$$

from Eq. 4.19, whose solution is

$$\lambda_\pm = \frac{-(en - gl) \pm \mathcal{R}}{2(fn - gm)}. \tag{6.34}$$

This comes from Eq. 4.20; $\mathcal{R}$, the discriminant of the quadratic equation in Eq. 6.33, is defined in Eq. 4.21 as

$$\mathcal{R} = (en - gl)^2 - 4(em - fl)(fn - gm). \tag{6.35}$$

An alternative expression for $\lambda_\pm$ is

$$\lambda_\pm = \frac{-2(em - fl)}{(en - gl) \pm \mathcal{R}}. \tag{6.36}$$

From Eqs. 4.22 and 4.23 the two orthogonal principal directions are given by the unit vectors

$$\mathbf{t}_\pm = \frac{\mathbf{W}_p + \lambda_\pm \mathbf{W}_q}{\sqrt{e + f\lambda_\pm + g\lambda_\pm^2}}. \tag{6.37}$$

Suppose next that the p and q directions are the principal directions. Then $\mathbf{W}_p$ and $\mathbf{W}_q$ are vectors tangent to the two orthogonal geodesic curves in the principal directions. It follows that $f = 0$, because the two directions are orthogonal, and that $m = 0$ since they are principal directions. From Eqs. 4.47 and 4.53

$$\begin{cases} E v_p v_q + F(v_p w_q + v_q w_p) + G w_p w_q &= 0 \\ L v_p v_q + M(v_p w_q + v_q w_p) + N w_p w_q &= 0. \end{cases} \tag{6.38}$$

If we make the substitutions $w_p = \lambda_p v_p$ and $w_q = \lambda_q v_q$, just as we had done in Chapter 4, this becomes

$$\begin{cases} E + F(\lambda_p + \lambda_q) + G\lambda_p\lambda_q &= 0 \\ L + M(\lambda_p + \lambda_q) + N\lambda_p\lambda_q &= 0. \end{cases} \tag{6.39}$$

Now we make the identification $\lambda_p = \lambda_+$ and $\lambda_q = \lambda_-$ where λ_+ and λ_- are solutions of the quadratic polynomial equation in Eq. 4.19. Their sum and product from one of Gauss' many theorems, given in Eq. 4.23, satisfy both equations of Eq. 6.39.

As in Chapter 4 when $f = m = 0$ the polynomial in Eq. 6.33 becomes ephemeral and we can take as the two roots $\lambda_p = 0$ and $\lambda_q = \infty$. We can justify this by noting that, in general $en - gl \neq 0$ (an exception to this will be discussed below) so that neither $en - gl + \mathcal{R}$ nor $(en - gl) - \mathcal{R}$ vanish. On the other hand $fl - em \equiv 0$. Now refer to Eq. 6.34, in which we can see that $\lambda_q = \infty$, and to Eq. 6.36 which yields $\lambda_p = 0$.

The remaining two equations in Eq. 4.45 become

$$\begin{cases} e = v_p^2(E + 2F\lambda_p + G\lambda_p^2) \\ g = v_q^2(E + 2F\lambda_q + G\lambda_q^2); \end{cases} \tag{6.40}$$

those from Eq. 4.52 are

$$\begin{cases} l \;= v_p^2(L + 2M\lambda_p + N\lambda_p^2) \\ n = v_q^2(L + 2M\lambda_q + N\lambda_q^2). \end{cases} \tag{6.41}$$

Now from Eq. 6.32 we can see that the two principal curvatures are

$$\frac{1}{\rho_p} = \frac{l}{e}, \qquad \frac{1}{\rho_q} = \frac{n}{g} \tag{6.42}$$

just as they should be.

Now we can return to the question of the case when $en - gl = 0$. This happens if and only if the two curvatures are equal in which case there are no principal directions and all three coefficients of the quadratic polynomial equation that defines the principal directions are *zero*. Such a point is called an *umbilical point.*

6.5 Asymptotic Curves and Isotropic Directions

In Chapter 4 we mentioned two features of geometric surfaces, asymptotic curves and isotropic directions. Asymptotic directions, as in the case of any direction on a surface, is given by the expression for the unit tangent vector, Eq. 4.14, where in this case λ is the solution of

$$L + 2M\lambda + N\lambda^2 = 0, \tag{6.43}$$

this from Eq. 4.12. If Eq. 6.43 has a real solution then, according to Eq. 4.15, the curvature of a geodesic curve in the asymptotic directions must vanish. The asymptotic curves then must be straight lines.

By substituting the second fundamental quantities from Eq. 6.11 this becomes

$$\begin{aligned} &[n^2u^2k_{vv} - q(u^2 + v^2)] \\ &\quad + 2[n^2u^2k_{vw} - qvw]\lambda \\ &\qquad + [n^2u^2k_{ww} - q(u^2 + w^2)]\lambda^2 = 0. \end{aligned} \tag{6.44}$$

The discriminant of this quadratic polynomial equation is

$$\begin{aligned}\Delta^2 &= [n^2u^2k_{vw} - qvw]^2 \\ &\quad - [n^2u^2k_{vv} - q(u^2+v^2)][n^2u^2k_{ww} - q(u^2+w^2)] \\ &= -\{n^4u^4(k_{vv}k_{ww} - k_{vw}^2) \\ &\quad - qn^2u^2[k_{vv}(n^2 - v^2) + k_{ww}(n^2 - w^2) - 2vwk_{vw}] \\ &\quad + q^2[u^4 + u^2(v^2 + w^2)]\} \\ &= -n^2u^2[q^2 - qH + n^2u^2T^2],\end{aligned} \tag{6.45}$$

where we have used Eqs. 6.6 and 6.7.

For λ to be real the discriminant must be positive so that

$$q^2 - Hq + n^2u^2T^2 = n^3uD = n^2\rho_+\rho_- \leq 0. \tag{6.46}$$

so that ρ_+ and ρ_- must have opposite signs. This means that the wavefront must lie in a region where it has a saddle point. For this to happen the wavefront must lie between the two caustic sheets which implies that these two branches of the caustic must be disjoint. This can be seen in what follows.

The values of q for which asymptotic directions exist are determined by the inequality in Eq. 6.46. The boundaries for these values must be the two solutions of the equation obtained by replacing the inequality by an equal sign thus

$$q^2 - Hq + n^2u^2T^2 = 0$$

where the discriminant is $H^2 - 4n^2u^2T^2 = \mathcal{S}^2$ from Eq. 6.15 so that the two values of $q_\pm$ are

$$q_\pm = \frac{1}{2}(H \pm \mathcal{S}) \tag{6.47}$$

But $q = ns - k + \mathbf{S} \cdot \mathbf{K}$ so that the bounds of s are

$$\overline{s}_\pm = \frac{1}{n}\left[k - (\mathbf{S} \cdot \mathbf{K}) + \frac{1}{2}(H \pm \mathcal{S})\right] \tag{6.48}$$

From Eq. 6.21 we can calculate

$$\mathbf{C}_\pm \cdot \mathbf{S} = \frac{1}{2}(H \pm \mathcal{S}) - (\mathbf{S} \cdot \mathbf{K})$$

from which we can get

$$\overline{s}_\pm = \frac{1}{n}(k + \mathbf{C}_\pm \cdot \mathbf{S}). \tag{6.49}$$

The bounds of s, $s_\pm$, are functions of v and w.

When we substitute the two values, $q_\pm$ into the expression for the wavefronts given in Eq. 5.72 we get

$$\mathbf{W}_\perp = \frac{1}{2n^2}(H \pm \mathcal{S})\mathbf{S} - \mathbf{K} = \mathbf{C}_+ \tag{6.50}$$

which shows that the boundary surfaces for the region in which asymptotic directions exist is exactly the two sheets of the caustic surface. The two sheets must therefore be disjoint which implies that H must vanish and that T^2 be negative.

What all of this means is that a condition of the existence of asymptotic curves on a wavefront is that

$$H = (n^2 - v^2)k_{vv} + (n^2 - w^2)k_{ww} - 2vwk_{vw} = 0, \tag{6.51}$$

a partial differential equation of the second order. This we will return to in Chapter 8, Aberrations in Finite Terms.

The isotropic directions are treated in a similar way. Here the defining equation is

$$E + 2F\lambda + G\lambda^2 = 0, \tag{6.52}$$

which becomes, after substitution from Eq. 6.5

$$\begin{aligned}
&[(q^2 - n^2u^2T^2)(n^2 - w^2) + n^2u^2(H - 2q)k_{vv}] \\
&\quad +[(q^2 - n^2u^2T^2)vw + n^2u^2(H - 2q)k_{vw}]\lambda \\
&\qquad +[(q^2 - n^2u^2T^2)(n^2 - v^2) + n^2u^2(H - 2q)k_{ww}]\lambda^2 = 0
\end{aligned} \tag{6.53}$$

whose discriminant is far more difficult to calculate and far less rewarding.

$$\begin{aligned}
\Delta^2 &= [(q^2 - n^2u^2T^2)vw + n^2u^2(H - 2q)k_{vw}]^2 \\
&\quad -[(q^2 - n^2u^2T^2)(n^2 - w^2) + n^2u^2(H - 2q)k_{vv}] \\
&\qquad \times[(q^2 - n^2u^2T^2)(n^2 - v^2) + n^2u^2(H - 2q)k_{ww}] \\
&= -n^2u^2\{(q^2 - n^2u^2T^2)^2 + H(q^2 - n^2u^2T^2)(H - 2q) \\
&\qquad +n^2u^2T^2(H - 2q)^2\} \\
&= -n^2u^2(q^2 - Hq + n^2u^2T^2)^2,
\end{aligned} \tag{6.54}$$

This quantity can never be positive so that there are no real solutions for λ. This indicates that there can be no isotropic directions for wavefronts.

This concludes this chapter on wavefronts. In it we have found the principal directions and principal curvatures of each wavefront in a wavefront train as well as its caustic surface. We derived expressions for transforming the wavefront parameters. We also looked at the possibility of there being asymptotic curves imbedded on the wavefront. We also found that wavefronts can have no isotropic directions. We examined two special cases, the plane wavefront and the train of spherical wavefronts. These items will be treated further in subsequent chapters.

At this point it should be evident that the k-function and its derivatives determine completely the structure of the individual wavefronts in a train and the associated caustic surface.

7 Ray Tracing: Generalized and Otherwise

Rays are the trajectories of those mysterious particles of light that, anthropomorphically, seek the quickest path in their peregrinations from one point to another. It is indeed remarkable that these mindless abstractions can determine their routes without the need of any of the tools, abaci or Crays, that we mere humans require. So we come to the point in our journey where we need to contemplate methods and aids to match wits with these peculiar objects.

Rays, those trajectories of non existent corpuscles, are ephemeral things with no visible means of support. Yet the concept is real enough to be eminently useful in the analysis and design of optical systems.

A lens is an array of light transmitting or reflecting elements, usually glass, each with a predetermined refractive index and dispersion, whose surfaces are ground and polished according to some design prescription. They can be separated by air, an optical medium, or they can be cemented together.

Ray tracing is the process whereby ray intercepts and direction cosines are calculated from surface to surface, through each medium in the optical system. The object of ray tracing is to determine whether the lens satisfies its prescribed specifications prior to its fabrication and is therefore of fundamental importance to the optical designer.

In interpreting the results of ray tracing several devices are used. Spot diagrams[1] are plots of the coordinates of rays, originating at some object point, that are intercepted by a plane in image space. Herzberger's diapoints[2], points formed by the intersection of rays with the meridional plane, can also be displayed. Plots of the caustics points are obtained from generalized ray tracing. Ray tracing also provides data for programs that attempt to predict the effects of diffraction on lens performance. These go beyond the scope of this work and will not be discussed here.

We have already encountered ray paths in inhomogeneous media in Chapters 1 and 2. In this chapter we will consider only homogeneous, isotropic media in which the refractive index is constant and rays are straight lines. First we will develop the mechanics of conventional ray tracing (referred to as *otherwise* in the chapter heading) where the parameters are the ray's direction cosines and the coordinates of a point on that ray.

Following this we will generalize the ray tracing equations to include the calculation of the local properties of the wavefront; its principal directions and its principal curvatures, at the point where ray and wavefront intersect. These are a generalization of the Coddington equations. I call this processes *generalized ray tracing*. With these it is possible to calculate

[1]Washer 1966, Stavroudis 1967.

[2]Welford 1974, Chapter 4.

The Mathematics of Geometrical and Physical Optics: The k-function and its Ramifications. O.N. Stavroudis

ISBN: 3-527-40448-1

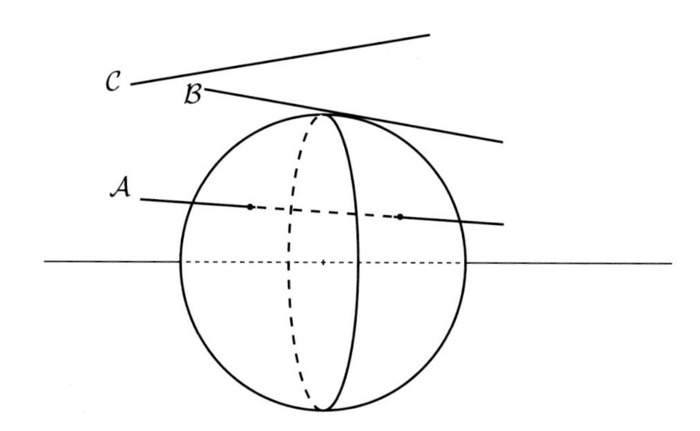

Figure 7.1: *Rays Intersecting a Sphere.* Ray $\mathcal{A}$ intercepts the surface in two places corresponding to the two roots of the quadratic equation. Ray $\mathcal{B}$ is tangent; the two roots are equal. Ray $\mathcal{C}$ misses the sphere and the solution is complex.

the caustic surfaces associated with some fixed object point and generated by the lens. These equations are singularly useful in dealing with systems that lack rotational symmetry.

7.1 The Transfer Equations

Lets begin with a ray that passes through some point $\mathbf{R}_0$ with its direction given by the reduced cosine vector $\mathbf{S} = (u,\ v,\ w)$, where $\mathbf{S}^2 = n^2$, as noted in Chapter 4. The point $\mathbf{R}_0$ may be where the ray intersects a preceding surface in this ray trace or is perhaps an object point. Let the next succeeding surface in an optical system be given by the equation

$$\mathcal{F}(x,\ y,\ z) = \mathcal{F}(\mathbf{R}) = 0. \tag{7.1}$$

Any point $\mathbf{R}_\lambda$ on the ray is given by the expression

$$\mathbf{R}_\lambda = \mathbf{R} + (\lambda/n)\mathbf{S}, \tag{7.2}$$

where n is the refractive index of the medium and where λ is the distance along the ray from $\mathbf{R}$.

In practice, the transfer operation is accompanied by a shift of the coordinate axes to the new surface; e.g., for a rotationally symmetric system $\mathbf{R}_\lambda = \mathbf{R} - t\mathbf{Z} + \lambda\ \mathbf{S}$, where $\mathbf{Z}$ is the unit vector along the axis of rotation and t is the distance to the next surface along this axis. By substituting $\mathbf{R}_\lambda$ given in Eq. 7.2 into the equation for the surface in Eq. 7.1 we obtain an equation for λ

$$\mathcal{F}(\mathbf{R}_\lambda) = \mathcal{F}(\mathbf{R} + (\lambda/n)\mathbf{S}) = 0, \tag{7.3}$$

whose solution gives values of λ which, when substituted back into Eq. 7.2, yield points of incidence. Figure 7.1 is an attempt to show this.

When we say *a* point of incidence we imply that there may be more than one; indeed, the existence of multiple points of intersection of the ray with a refracting or reflecting surface is

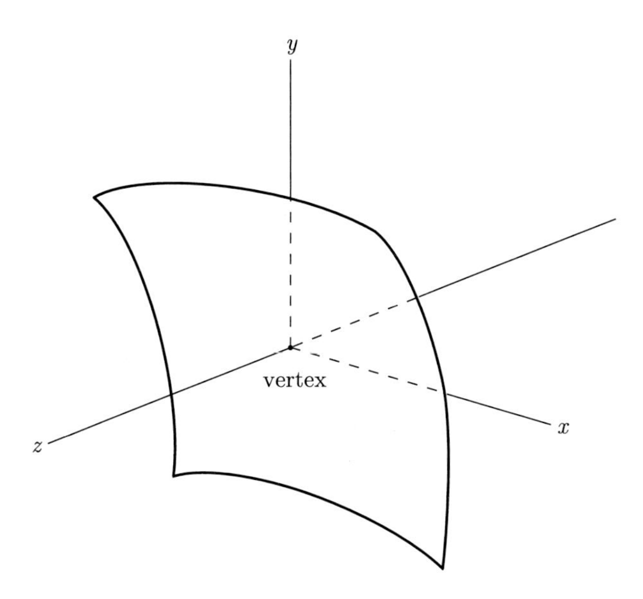

Figure 7.2: *The Local Coordinate System.* The origin is placed at the vertex of the rotationally symmetric refracting surface – the point where the lens axis intersects the surface.

one of the annoying difficulties in developing formulas for the transfer operation, particularly for systems that lack rotational symmetry. The commonest refracting surface is the sphere, a quadratic surface where the function $\mathcal{F}$ is a polynomial of the second degree. The equation corresponding to Eq. 7.3 will have two roots that identify the two points of intersection of a straight line with that sphere. If these roots are equal then the ray is tangent to the refracting sphere; if they are complex then the ray has no real point of intersection and misses the surface. What is said here applies to any quadratic surface and therefore includes conic surfaces such as prolate spheroids, oblate spheroids, paraboloids and hyperboloids of one and two sheets.

In rotationally symmetric systems, the local coordinate origin is taken to be at the intersection of the system's axis of symmetry with the refracting conic surface, usually referred to as the *vertex*, and is shown in Fig. 7.2. The point of incidence is that branch of the quadratic solution nearest the origin. This way of avoiding decision making fails for systems that are off axis or not rotationally symmetric, especially when the details of the ray tracing calculations are obscured by the *arcana* of a computer program. This problem will be discussed in Chapter 11 that deals with the modern schiefspiegler, a wildly off-axis reflecting optical system.

The transfer operation does more than find the points of incidence but includes the calculation of other quantities as well. The most important of these is the unit normal vector of the refracting surface at the point of incidence that we encountered in Chapter 4

$$\mathbf{N} = \frac{\nabla\mathcal{F}}{\sqrt{(\nabla\mathcal{F})^2}}. \tag{4.98}$$

For generalized ray tracing, the principal directions and the principal curvatures of the wavefront at the point of intersection of ray and wavefront will also be required.

7.2 The Ancillary Quantities

The details of calculating **N** come next. The equation in Eq. 7.1 defines z as an implicit function of x and y so that its partial derivatives become

$$\frac{\partial \mathcal{F}}{\partial x} + \frac{\partial \mathcal{F}}{\partial z}\frac{\partial z}{\partial x} = 0, \qquad \frac{\partial \mathcal{F}}{\partial y} + \frac{\partial \mathcal{F}}{\partial z}\frac{\partial z}{\partial y} = 0. \tag{7.4}$$

The gradient then becomes

$$\nabla \mathcal{F} = \left(\frac{\partial \mathcal{F}}{\partial x}, \frac{\partial \mathcal{F}}{\partial y}, \frac{\partial \mathcal{F}}{\partial z}\right) = -\frac{\partial \mathcal{F}}{\partial z}\left(\frac{\partial z}{\partial x}, \frac{\partial z}{\partial y}, -1\right). \tag{7.5}$$

We will also need to calculate the surface parameters of the refracting surface at the point of incidence, the principal directions and curvatures, that are required for generalized ray tracing.

In Chapter 4 we saw that a surface can be represented as a vector function of two parameters and that those two parameters can be the x and y coordinates of a point on a plane. Then the vector function can be written as

$$\text{(4.90)} \qquad \mathbf{R}(x, y) = \big(x, y, z(x, y)\big),$$

so that the partial derivatives are

$$\text{(4.91)} \qquad \mathbf{R}_x = (1, 0, z_x) \qquad \mathbf{R}_y = (0, 1, z_y),$$

from which we can calculate the unit normal vector

$$\text{(4.93)} \qquad \mathbf{N} = -\frac{(z_x, z_y, -1)}{\sqrt{1 + z_x^2 + z_y^2}}$$

and the first fundamental quantities from Eq. 4.4

$$E = 1 + z_x^2, \qquad F = z_x z_y, \qquad G = 1 + z_y^2. \tag{7.6}$$

Also using Eq. 4.11 we can calculate the second fundamental quantities

$$L = \frac{z_{xx}}{\sqrt{1 + z_x^2 + z_y^2}} \qquad M = \frac{z_{xy}}{\sqrt{1 + z_x^2 + z_y^2}} \qquad N = \frac{z_{yy}}{\sqrt{1 + z_x^2 + z_y^2}}. \tag{7.7}$$

From Eqs. 7.6 and 7.7, using Eqs. 4.18, 4.19, 4.21 and 4.25, the principal directions and principal curvatures of the refracting surfaces at the point of incidence can be calculated. These quantities will be needed when we come to generalized ray tracing later in this chapter.

7.3 The Refraction Equations

In Chapter 3 we derived the two versions of Snell's law, the vector form

$$\text{(3.51)} \qquad \mathbf{S_1}{\times}\mathbf{N} = \mathbf{S}{\times}\mathbf{N},$$

and the scalar form

$$(3.49) \qquad n_1 \sin r = n \sin i,$$

where i stands for the angle of incidence, the angle between the surface normal and the incident ray, and r, the angle of refraction, the angle between the surface normal and the refracted ray, as discussed in Chapter 4.

The two *reduced* ray vectors are $\mathbf{S}$ for the incident ray and $\mathbf{S}_1$ for the refracted ray so that $n \cos i = \mathbf{S} \cdot \mathbf{N}$ and $n_1 \cos r = \mathbf{S}_1 \cdot \mathbf{N}$ where $\mathbf{N}$ is the unit vector normal to the refracting surface at the point of incidence as in Eq. 4.93.

Now we rearrange the vector form of Snell's law, Eq. 3.51, as follows:

$$(\mathbf{S}_1 - \mathbf{S}) \times \mathbf{N} = 0, \tag{7.8}$$

which shows that the vector $\mathbf{S}_1 - \mathbf{S}$ must be parallel to $\mathbf{N}$. Then there must exist some quantity γ so that $\mathbf{S}_1 - \mathbf{S} = \gamma \mathbf{N}$ whence comes

$$\mathbf{S}_1 = \mathbf{S} + \gamma \mathbf{N}, \tag{7.9}$$

the refraction equation. By taking the scalar product of this with $\mathbf{N}$ we get

$$\gamma = \mathbf{S}_1 \cdot \mathbf{N} - \mathbf{S} \cdot \mathbf{N} = n_1 \cos r - n \cos i. \tag{7.10}$$

The formula for reflection is highly artificial. In the usual convention one sets $n_1 = -n$ ($n = 1$ in air) in Eqs. 7.9 and 7.10. Since $\sin r = -\sin i$ it must be that $\cos r = \cos i$ so that, $\gamma = -2 \cos i$ and,

$$\mathbf{S}_1 = -\mathbf{S} + 2 \cos i \, \mathbf{N}. \tag{7.11}$$

So with the two equations, Eqs. 7.9 and 7.10, we are able to find the refracted ray vector $\mathbf{S}_1$ when the incident ray vector $\mathbf{S}$, the surface normal $\mathbf{N}$ and refractive indices n and n_1 are known. Equation 3.49, the scalar version of Snell's law, is used to find $\sin r$ from $\sin i$ and the two refractive indices.

From Eq. 7.9 it is clear that $\mathbf{S}_1$ is a linear combination of $\mathbf{N}$ and $\mathbf{S}$. The three vectors are then coplanar; the plane so determined is called the *plane of incidence*.

This completes this section on ordinary ray tracing. With the formulas derived here one can trace a system of rays through an optical system to some plane in image space where the ray intercepts can be plotted. The result is a *spot diagram*[3] (once called a *focal plot* in the ancient literature) that can illustrate quite graphically the distribution of the light on some plane in image space. With ray tracing, graphs can be plotted of the point coordinates in image space versus coordinates on an entrance pupil to reveal details of the aberrations associated with a particular object point. When image point coordinates are plotted versus object point coordinates the data will show distortion and image curvature.[4] These are details that we will not delve into here.

---

[3] Stavroudis and Sutton 1965.

[4] Stavroudis, 1972a, Chapter XII.

One needs to add to all this that the spot diagram is the geometric analog of the star image test in which collimated light is made to enter the lens under test. The quality of the image formed and the distribution of the light in that image provides measures of the lenses performance. Because of this, the spot diagram is able to indicate how a lens will behave before its fabrication.[5]

But there is a limitation. We speak of *geometric* aberrations. When these are large they dominate image formation. But if they are sufficiently small, diffraction effects dominate and geometry has little or no influence of the formation of an image. Such a situation is commonly referred to as being *diffraction limited*.[6]

7.4 Rotational Symmetry

By far most optical systems are rotationally symmetric; each surface in such a system must itself be rotationally symmetric and then all of the lines of symmetry of the component surfaces must coincide, thus forming an *axis of symmetry* for the entire system.

As a matter of convenience we will use a coordinate system in which the z-axis is placed on the axis of symmetry and a local coordinate origin is located where this axis and the refracting surface intersect. This point is usually referred to as the *vertex*.

In a rotationally symmetric system the axis of symmetry and an object point determine a plane, the *meridional plane*. Rays on this plane are called *meridional rays*; rays that are not meridional rays are called *skew rays*, thus distinguishing two disjoint classes of rays. Meridional rays remain meridional rays as they progress through a rotationally symmetric optical system; skew rays can never become meridional rays. Of course, this distinction doesn't apply to off-axis systems.

In rotationally symmetric systems *diapoints*[7] can be calculated using the ray tracing equations; these are the points of intersection of skew rays on a meridional plane and are useful in visualizing the geometric aberrations of a lens. Diapoints and their uses will be discussed further in Chapter 8, Aberrations in Finite Terms.

One additional point. If an optical system is rotationally symmetric then there is a quantity called *skewness* that is an invariant under both refraction and transfer. If the axis of symmetry is the z-axis, if $\mathbf{P} = (x,\ y,\ z)$ is any point on a skew ray with a direction cosine vector is $\mathbf{S} = (\xi,\ \eta,\ \zeta)$ then skewness is given by[8]

$$K = n(x\eta - y\xi) = n\,\mathbf{Z}\cdot(\mathbf{P}\times\mathbf{S}). \tag{7.12}$$

The skewness K is a measure of the angle a skew ray makes with the meridional plane; it vanishes for meridional rays. That it is indeed invariant can be verified from Eqs. 7.2 and 7.9.

[5] Stavroudis 1972a, pp. 228 et seq.
[6] Born and Wolf 1998, Chapter XII.
[7] Herzberger 1958, Chapter 7.
[8] Stavroudis 1972a, pp. 208–209.

7.5 The Paraxial Approximation

In rotationally symmetric systems it is common to construct linear approximations of the ray tracing equation in which all points are assumed to lie arbitrarily near the axis of symmetry. All quantities of quadratic degree and higher are small and are assumed to be *zero*. This neighborhood of the axis is referred to as the *paraxial region*; the resulting approximations are called the *paraxial ray tracing equations*. Ray traced with these approximate formulas are called *paraxial rays* even if their coordinates lie far outside the paraxial region.[9]

In what follows each surface is rotationally symmetric and is expressed in terms of a coordinate system whose origin is at its vertex, the point where the surface and its axis intersect. A power series expansion for z defines the surface,

$$z = \sum_{n=0}^{\infty} a_n (x^2 + y^2)^n$$

so that its two first partial derivatives are

$$\begin{aligned} z_x &= 2a_1 x + 2x \sum_{n=2}^{\infty} n a_n (x^2 + y^2)^{n-1} \\ z_y &= 2a_1 y + 2y \sum_{n=2}^{\infty} n a_n (x^2 + y^2)^{n-1}. \end{aligned}$$

Now referring to Eq. 4.93, in the paraxial approximation, ignoring all terms of degree greater than *one* the unit normal vector to the refracting surface becomes

$$\mathbf{N} = (x,\ y,\ 1). \tag{7.13}$$

The sine and cosine functions can also be represented as power series;

$$\sin u = u + \sum_{n=1}^{\infty} \frac{(-1)^n}{(2n+1)!} u^{2n+1} \qquad \cos u = 1 + \sum_{n=1}^{\infty} \frac{(-1)^n}{(2n)!} u^{2n}$$

so that, in the paraxial approximation, $\sin u = u$ and $\cos u = 1$. The paraxial approximation of the scalar form of Snell's law, Eq. 3.49, becomes

$$n_1 i_1 = n\, i, \tag{7.14}$$

Where i and i_1 are the paraxial angles of incidence and refraction. The refraction equation, Eqs. 7.9 and 7.10 degenerates to

$$n_1 u_1 = n u + c h (n_1 - n), \tag{7.15}$$

where c is the curvature of the refracting sphere and h is the height of the paraxial ray on the reflecting sphere. I won't attempt to explain how the paraxial approximation of a sphere is a plane and at the same time have a curvature.

[9] Stavroudis 2001, pp. 22–28.

The transfer equation, Eq. 7.2 becomes

$$h_1 = h + tu \tag{7.16}$$

where t is the distance along the axis of symmetry between the two surfaces, h is the paraxial ray hight on the initial surface and h_1 is its counterpart on the second surface.

The skewness invariant, given in Eq. 7.12 is a quadratic quantity and therefore vanishes under the paraxial approximation. This implies that paraxial rays cannot be skew rays; that all paraxial rays throughout an optical system must lie on a meridional plane.

The *Lagrange invariant* $\mathcal{L}$[10] is defined by

$$\mathcal{L} = n(y\overline{u} - \overline{y}u) = n_1(y_1\overline{u}_1 - \overline{y}_1u_1) \tag{7.17}$$

where $y,\ u$ and $\overline{y},\ \overline{u}$ are two paraxial rays. If $\mathcal{L} \neq 0$ the two rays are said to be independent.

If we think of paraxial rays as vectors on the meridian plane, since this plane is two-dimensional, then all paraxial rays can be written as a linear combination of only two vectors that can serve as a basis for a vector space consisting of the totality of the paraxial rays. The two base vectors must be independent; this is assured by the non vanishing of the Lagrangian invariant.

Suppose the two basis rays are h, for *height*, and u, for *slope* and $\overline{h}$ and $\overline{u}$; if the Lagrange invariant, does not vanish then these two rays are independent and can act as basis elements. The choice of these rays is entirely arbitrary; any pair of paraxial rays for which $\mathcal{L} \neq 0$ will do.

It is convenient to choose these basis rays to be the *paraxial marginal ray*, that ray that just clears the edges of all elements in an optical system, and the *paraxial chief ray*, the ray that passes through the center of the dominant aperture.[11]

These rays are fundamental to the calculation of the Seidel and higher order aberrations of a lens. This will not be discussed here.

7.6 Generalized Ray Tracing – Transfer

Generalized Ray Tracing[12] (generalized *Coddington Equations* might be a more accurate term) is the subject of these next sections. We are already accustomed to using the vector function of three components $\mathbf{W}(v,\ w,\ s)$ to represent a wavefront train as well as the use of $\mathbf{N}$ as the unit normal vector. To simplify our calculations assume that the parametric curves of the wavefronts are in the principle directions. In Chapter 4, using Eqs. 4.67 through 4.70, we saw that we could always make such a transformation of parameters. Take two positions of a wavefront corresponding to s and $s + \lambda$. Then

$$\mathbf{W}(v,\ w,\ s+\lambda) = \mathbf{W}(v,\ w,\ s) + \lambda\mathbf{N}(v,\ w,\ s),$$

or, to simplify matters

$$\mathbf{W}^{\lambda} = \mathbf{W} + \lambda\mathbf{N}.$$

[10] Stavroudis 1972a, pp. 209–210.
[11] Welford 1974, pp. 22–25.
[12] Stavroudis and Sutton, 1965.

Taking derivatives we get

$$\begin{cases} \mathbf{W}_v^\lambda = \mathbf{W}_v + \lambda \mathbf{N}_v = [1 - \lambda (L/E)] \mathbf{W}_v \\ \mathbf{W}_w^\lambda = \mathbf{W}_w + \lambda \mathbf{N}_w = [1 - \lambda (N/G)] \mathbf{W}_v \end{cases} \tag{7.18}$$

where we have used the Weingarten equations from Eq. 4.35. Here $F = 0$ and $M = 0$ because of our assumption that the lines of curvature were in the principal directions. These lead to the first fundamental quantities

$$E^\lambda = [1 - \lambda (L/E)]^2 E \qquad G^\lambda = [1 - \lambda (N/G)]^2 G, \tag{7.19}$$

and to the second fundamental quantities

$$L^\lambda = [1 - \lambda (L/E)] L \qquad N^\lambda = [1 - \lambda (N/G)] N, \tag{7.20}$$

obtained by again using the Weingarten equations. Next we use Eqs. 4.37 and 4.40 to calculate the principal curvatures

$$\frac{1}{\rho_v^\lambda} = \frac{L^\lambda}{E^\lambda} = \frac{1}{\rho_v - \lambda} \qquad \frac{1}{\rho_w^\lambda} = \frac{N^\lambda}{G^\lambda} = \frac{1}{\rho_w - \lambda}. \tag{7.21}$$

This tells us that the wavefront principal centers of curvature do not change as the wavefront progresses. We also can show that the principal directions do not change on transfer.

7.7 Generalized Ray Tracing – Preliminary Calculations

Recall that the plane of incidence is determined by the surface normal, $\mathbf{N}$, and the incident ray vector $\mathbf{S}$ and that as a consequence of Snell's law, Eq. 3.49, the refracted ray vector $\mathbf{S}_1$ also lies on this plane. We define the unit vector $\mathbf{P}$ to be perpendicular to the plane of incidence and use the ratio of the two expressions in Eqs. 3.49 and 3.51 to obtain

$$\mathbf{P} = \frac{\mathbf{S}_1 \times \mathbf{N}}{n \sin r} = \frac{\mathbf{S} \times \mathbf{N}}{n_1 \sin i}. \tag{7.22}$$

To show that these vectors are indeed equal both the vector form and the scalar form of Snell's law are used.

Next we define the vector $\mathbf{Q}$ for the incident wavefront, $\overline{\mathbf{Q}}$ for the refracting surface, and $\mathbf{Q}_1$ for the refracted wavefront as follows

$$\mathbf{Q} = (\mathbf{S} \times \mathbf{P})/n, \qquad \overline{\mathbf{Q}} = \mathbf{N} \times \mathbf{P}, \qquad \mathbf{Q}_1 = (\mathbf{S}_1 \times \mathbf{P})/n_1, \tag{7.23}$$

which leads to the following vector relationships

$$\begin{aligned} &\mathbf{P} = (\mathbf{Q} \times \mathbf{S})/n, \qquad &&\mathbf{P} = \overline{\mathbf{Q}} \times \mathbf{N}, \qquad &&\mathbf{P} = (\mathbf{Q}_1 \times \mathbf{S}_1)/n_1, \\ &\mathbf{S} = n(\mathbf{P} \times \mathbf{Q}), \qquad &&\mathbf{N} = \mathbf{P} \times \overline{\mathbf{Q}}, \qquad &&\mathbf{S}_1 = n_1(\mathbf{P} \times \mathbf{Q}_1). \end{aligned} \tag{7.24}$$

It follows from the definition of the $\mathbf{Q}$ vectors in Eq. 7.23 and the refraction equation in Eq. 7.9 that

$$n_1\mathbf{Q}_1 = n\mathbf{Q} + \gamma\overline{\mathbf{Q}}, \tag{7.25}$$

where γ is defined in Eq. 7.10. As a result of the transfer operations we have the point of incidence $\overline{\mathbf{P}}$, the two principal curvatures of the wavefront at $\overline{\mathbf{P}}$, $1/\rho_\xi$ and $1/\rho_\eta$, and one of the principal directions which we refer to as $\mathbf{T}$.

On the refracting surface at the point of incidence $\overline{\mathbf{R}}$, we do the same calculations and get the unit normal vector $\mathbf{N}$, the two principal curvatures $1/\overline{\rho}_\xi$ and $1/\overline{\rho}_\eta$, and one of the principal directions $\overline{\mathbf{T}}$.

For the incident wavefront we find the angle between the wavefront principal direction and $\mathbf{P}$

$$\cos\theta = \mathbf{T}\cdot\mathbf{P}. \tag{7.26}$$

Using Eq. 4.66 derived in Chapter 4 we find the curvatures of the geodesics tangent to $\mathbf{P}$ and $\mathbf{Q}$ and the torsion $1/\sigma$

$$\begin{cases} \dfrac{1}{\rho_p} = \dfrac{\cos^2\theta}{\rho_\xi} + \dfrac{\sin^2\theta}{\rho_\eta} \\ \dfrac{1}{\rho_q} = \dfrac{\sin^2\theta}{\rho_\xi} + \dfrac{\cos^2\theta}{\rho_\eta} \\ \dfrac{1}{\sigma} = \left(\dfrac{1}{\rho_\xi} - \dfrac{1}{\rho_\eta}\right)\sin\theta\cos\theta. \end{cases} \tag{7.27}$$

We make the same calculations for the refracting surface. First the angle between a principal direction of the refracting surface and $\mathbf{P}$

$$\cos\overline{\theta} = \overline{\mathbf{T}}\cdot\mathbf{P}, \tag{7.28}$$

then the curvatures of the geodesics tangent to $\mathbf{P}$ and $\mathbf{Q}$ and the torsion of the curve $1/\overline{\sigma}$

$$\begin{cases} \dfrac{1}{\overline{\rho}_p} = \dfrac{\cos^2\overline{\theta}}{\overline{\rho}_\xi} + \dfrac{\sin^2\overline{\theta}}{\overline{\rho}_\eta} \\ \dfrac{1}{\overline{\rho}_q} = \dfrac{\sin^2\overline{\theta}}{\overline{\rho}_\xi} + \dfrac{\cos^2\overline{\theta}}{\overline{\rho}_\eta} \\ \dfrac{1}{\overline{\sigma}} = \left(\dfrac{1}{\overline{\rho}_\xi} - \dfrac{1}{\overline{\rho}_\eta}\right)\sin\overline{\theta}\cos\overline{\theta}, \end{cases} \tag{7.29}$$

exactly as was done in Eq. 4.66.

Recall the Frenet-Serret equations and the relations between $\mathbf{t}$, $\mathbf{n}$ and $\mathbf{b}$ from Chapter 2

$$\text{(2.10)} \qquad \mathbf{t} = \mathbf{n}\times\mathbf{b}, \qquad \mathbf{n} = \mathbf{b}\times\mathbf{t}, \qquad \mathbf{b} = \mathbf{t}\times\mathbf{n},$$

and

$$(2.20) \quad \begin{cases} (\mathbf{t}\cdot\nabla)\mathbf{t} = & \frac{1}{\rho}\mathbf{n} \\ (\mathbf{t}\cdot\nabla)\mathbf{n} = -\frac{1}{\rho}\mathbf{t} & +\frac{1}{\tau}\mathbf{b} \\ (\mathbf{t}\cdot\nabla)\mathbf{b} = & -\frac{1}{\tau}\mathbf{n} \end{cases}$$

where we have used directional derivatives.

For the geodesic curve tangent to $\mathbf{P}$ we make the following identifications:

$$\begin{cases} \mathbf{t}_p = \mathbf{P} \\ \mathbf{n}_p = \mathbf{S}/n \\ \mathbf{b}_p = \mathbf{t}_p \times \mathbf{n}_p = (\mathbf{P}\times\mathbf{S})/n = -\mathbf{Q} \end{cases} \tag{7.30}$$

where we have made use of Eqs. 7.23 and 7.24.

Applying these relations to the Frenet-Serret equations given in Eq. 2.12, results in

$$(2.12) \quad \begin{cases} (\mathbf{t}_p\cdot\nabla)\mathbf{t}_p = & \frac{1}{n\rho_p}\mathbf{n}_p \\ (\mathbf{t}_p\cdot\nabla)\mathbf{n}_p = -\frac{n}{\rho_p}\mathbf{t}_p & +\frac{n}{\tau_p}\mathbf{b}_p \\ (\mathbf{t}_p\cdot\nabla)\mathbf{b}_p = & -\frac{1}{n\tau_p}\mathbf{n}_p \end{cases}$$

which becomes, using Eq. 7.30

$$\begin{cases} (\mathbf{P}\cdot\nabla)\mathbf{P} = & \frac{1}{n\rho_p}\mathbf{S} \\ (\mathbf{P}\cdot\nabla)\mathbf{S} = -\frac{n}{\rho_p}\mathbf{P} & +\frac{n}{\tau_p}\mathbf{Q} \\ -(\mathbf{P}\cdot\nabla)\mathbf{Q} = & -\frac{1}{n\tau_p}\mathbf{S}. \end{cases} \tag{7.31}$$

Now for the geodesic curve tangent to $\mathbf{Q}$ we make these identifications again using Eqs. 7.23 and 7.24

$$\begin{cases} \mathbf{t}_q = \mathbf{Q} \\ \mathbf{n}_q = \mathbf{S}/n \\ \mathbf{b}_q = \mathbf{t}_q \times \mathbf{n}_q = (\mathbf{Q}\times\mathbf{S})/n = \mathbf{P}. \end{cases} \tag{7.32}$$

By making the obvious substitutions from Eq. 7.32, as in the previous case we get

$$\begin{cases} (\mathbf{Q}\cdot\nabla)\mathbf{Q} = \dfrac{1}{n\rho_q}\mathbf{S} \\ (\mathbf{Q}\cdot\nabla)\mathbf{S} = -\dfrac{n}{\rho_q}\mathbf{Q} + \dfrac{n}{\tau_q}\mathbf{P} \\ (\mathbf{Q}\cdot\nabla)\mathbf{P} = -\dfrac{1}{n\tau_q}\mathbf{S}. \end{cases} \tag{7.33}$$

We know from Eq. 4.41 that the two values of torsion are equal in magnitude and opposite in sign so that we may define

$$\frac{1}{\sigma} = -\frac{1}{\tau_p} = \frac{1}{\tau_q}. \tag{7.34}$$

This results in, for the **P** geodesic, from Eq. 7.31

$$\begin{cases} (\mathbf{P}\cdot\nabla)\mathbf{P} = \dfrac{1}{n\rho_p}\mathbf{S} \\ (\mathbf{P}\cdot\nabla)\mathbf{S} = -\dfrac{n}{\rho_p}\mathbf{P} + \dfrac{n}{\sigma}\mathbf{Q} \\ (\mathbf{P}\cdot\nabla)\mathbf{Q} = -\dfrac{1}{n\sigma}\mathbf{S}, \end{cases} \tag{7.35}$$

and for the **Q** geodesic, where we use Eq. 7.33

$$\begin{cases} (\mathbf{Q}\cdot\nabla)\mathbf{Q} = \dfrac{1}{n\rho_q}\mathbf{S} \\ (\mathbf{Q}\cdot\nabla)\mathbf{S} = -\dfrac{n}{\rho_q}\mathbf{Q} + \dfrac{n}{\sigma}\mathbf{P} \\ (\mathbf{P}\cdot\nabla)\mathbf{P} = -\dfrac{1}{n\sigma}\mathbf{S}. \end{cases} \tag{7.36}$$

Since **S** and **Q**, **N** and $\overline{\mathbf{Q}}$, $\mathbf{S}_1$ and $\mathbf{Q}_1$ are all coplanar, we may also write

$$\begin{aligned} \mathbf{S} &= \mathbf{N}\cos i + \overline{\mathbf{Q}}\sin i, & \mathbf{S}_1 &= \mathbf{N}\cos r + \overline{\mathbf{Q}}\sin r, \\ \mathbf{Q} &= -\mathbf{N}\sin i + \overline{\mathbf{Q}}\cos i, & \mathbf{Q}_1 &= -\mathbf{N}\sin r + \overline{\mathbf{Q}}\cos r, \end{aligned} \tag{7.37}$$

from which we can get

$$\overline{\mathbf{Q}} = \mathbf{S}\sin i + \mathbf{Q}\cos i = \mathbf{S}_1\sin r + \mathbf{Q}_1\cos r. \tag{7.38}$$

Finally we note that, since rays in homogeneous media are always straight lines

$$(\mathbf{S}\cdot\nabla)\mathbf{S} = (\mathbf{S}_1\cdot\nabla)\mathbf{S}_1 = 0. \tag{7.39}$$

7.8 Generalized Ray Tracing – Refraction

Now we come to the derivation of the equations for generalized ray tracing[13]. We take the derivative of the vector form of Snell's law, given in Eq. 3.51, in the direction of $\mathbf{P}$ and get

$$\begin{aligned}&[(\mathbf{P}\cdot\nabla)\mathbf{S}_1]\times\mathbf{N}+\mathbf{S}_1\times[(\mathbf{P}\cdot\nabla)\mathbf{N}]\\&\quad=[(\mathbf{P}\cdot\nabla)\mathbf{S}]\times\mathbf{N}+\mathbf{S}\times[(\mathbf{P}\cdot\nabla)\mathbf{N}],\end{aligned}\tag{7.40}$$

The process of evaluating this is long and complicated so we must calculate separately the individual terms of Eq. 7.40 as follows.

$$\begin{aligned}[(\mathbf{P}\cdot\nabla)\mathbf{S}_1]\times\mathbf{N}&=\left(-\frac{n_1}{\rho_p'}\mathbf{P}+\frac{n_1}{\sigma'}\mathbf{Q}_1\right)\times\mathbf{N}\\&=-\frac{n_1}{\rho_p'}(\mathbf{P}\times\mathbf{N})+\frac{n_1}{\sigma'}(-\mathbf{N}\sin i+\overline{\mathbf{Q}}\cos i)\times\mathbf{N}\\&=-\frac{n_1}{\rho_p'}\overline{\mathbf{Q}}+\frac{n_1}{\sigma'}\mathbf{P}.\end{aligned}\tag{7.41}$$

Here and in the following calculations we have used Eqs. 7.23, 7.24, 7.35 and 7.38

$$\begin{aligned}[(\mathbf{P}\cdot\nabla)\mathbf{S}]\times\mathbf{N}&=\left(-\frac{n}{\rho_p}\mathbf{P}+\frac{n}{\sigma}\mathbf{Q}\right)\times\mathbf{N}\\&=-\frac{n}{\rho_p}(\mathbf{P}\times\mathbf{N})+\frac{n}{\sigma}(-\mathbf{N}\sin i+\overline{\mathbf{Q}}\cos i)\times\mathbf{N}\\&=\frac{n}{\rho_p}\overline{\mathbf{Q}}+\frac{n}{\sigma}\mathbf{P}.\end{aligned}\tag{7.42}$$

$$\begin{aligned}\mathbf{S}_1\times[(\mathbf{P}\cdot\nabla)\mathbf{N}]&=\mathbf{S}_1\times\left[-\frac{1}{\bar{\rho}_p}\mathbf{P}+\frac{1}{\bar{\sigma}}\overline{\mathbf{Q}}\right]\\&=-\frac{1}{\bar{\rho}_p}\mathbf{S}_1\times\mathbf{P}+\frac{n_1}{\bar{\sigma}}(\mathbf{N}\cos r+\overline{\mathbf{Q}}\cos r)\times\overline{\mathbf{Q}}\\&=-\frac{n_1}{\bar{\rho}_p}\mathbf{Q}_1-\frac{n_1\cos r}{\bar{\sigma}}\mathbf{P}.\end{aligned}\tag{7.43}$$

$$\begin{aligned}\mathbf{S}\times[(\mathbf{P}\cdot\nabla)\mathbf{N}]&=\mathbf{S}\times\left[-\frac{1}{\bar{\rho}_p}\mathbf{P}+\frac{1}{\bar{\sigma}}\overline{\mathbf{Q}}\right]\\&=-\frac{1}{\bar{\rho}_p}\mathbf{S}\times\mathbf{P}+\frac{n}{\bar{\sigma}}(\mathbf{N}\cos i+\overline{\mathbf{Q}}\sin i)\times\overline{\mathbf{Q}}\\&=-\frac{n}{\bar{\rho}_p}\mathbf{Q}-\frac{n\cos r}{\bar{\sigma}}\mathbf{P}.\end{aligned}\tag{7.44}$$

[13] Stavroudis 1976.

Putting these, Eqs. 7.41, 7.42, 7.43 and 7.44, all together yields

$$\begin{aligned}
&\frac{n_1}{\rho'_p}\overline{\mathbf{Q}}+\frac{n_1\cos r}{\sigma'}\mathbf{P}-\frac{n_1\cos r}{\bar{\sigma}}\mathbf{P}-\frac{1}{\bar{\rho}_p}(n\mathbf{Q}+\gamma\overline{\mathbf{Q}})\\
&\quad=\frac{n}{\rho_p}\overline{\mathbf{Q}}+\frac{n\cos i}{\sigma}\mathbf{P}-\frac{n\cos i}{\bar{\sigma}}\mathbf{P}-\frac{n}{\bar{\rho}_p}\mathbf{Q}.
\end{aligned} \tag{7.45}$$

Collecting terms in $\mathbf{P}$ and $\overline{\mathbf{Q}}$ we get

$$\mathbf{P}\left[\frac{n_1\cos r}{\sigma'}-\frac{n_1\cos r}{\bar{\sigma}}-\frac{n\cos i}{\sigma}+\frac{n\cos i}{\bar{\sigma}}\right]+\overline{\mathbf{Q}}\left[\frac{n_1}{\rho'_p}-\frac{\gamma}{\bar{\rho}_p}-\frac{n}{\rho_p}\right]=0. \tag{7.46}$$

Since $\mathbf{P}$ and $\overline{\mathbf{Q}}$ are orthogonal each of the two coefficients must equal zero independently. This results in

$$\begin{cases}
\frac{n_1}{\rho'_p} &= \dfrac{n}{\rho_p}+\dfrac{\gamma}{\bar{\rho}_p}\\
\frac{n_1\cos r}{\sigma'} &= \dfrac{n\cos i}{\sigma}+\dfrac{n_1\cos r-n\cos i}{\bar{\sigma}}\\
&= \dfrac{n\cos i}{\sigma}+\dfrac{\gamma}{\bar{\sigma}}
\end{cases} \tag{7.47}$$

Now we take the derivative of the vector form of Snell's law in Eq. 3.50 in the direction of $\overline{\mathbf{Q}}$ and get

$$\begin{aligned}
&n_1\{[(\overline{\mathbf{Q}}\cdot\nabla)\mathbf{S}_1]\times\mathbf{N}+\mathbf{S}_1\times[(\overline{\mathbf{Q}}\cdot\nabla)\mathbf{N}]\}\\
&\quad=n\{[(\overline{\mathbf{Q}}\cdot\nabla)\mathbf{S}]\times\mathbf{N}+\mathbf{S}\times[(\overline{\mathbf{Q}}\cdot\nabla)\mathbf{N}]\},
\end{aligned} \tag{7.48}$$

As before we calculate the individual terms. Here we use Eq. 7.39 in addition to Eqs. 7.22, 7.23 and 7.36.

$$\begin{aligned}
(\overline{\mathbf{Q}}\cdot\nabla)\mathbf{S}_1&=[(\mathbf{S}_1\sin r+\mathbf{Q}_1\cos r)\cdot\nabla]\mathbf{S}_1\\
&=\cos r(\mathbf{Q}_1\cdot\nabla)\mathbf{S}_1\\
&=n\cos r\left[-\frac{1}{\rho'_q}\mathbf{Q}_1+\frac{1}{\sigma'}\mathbf{P}\right].
\end{aligned} \tag{7.49}$$

$$\begin{aligned}
[(\overline{\mathbf{Q}}\cdot\nabla)\mathbf{S}_1]\times\mathbf{N}&=n_1\cos r\left[\frac{-1}{\rho'_q}\mathbf{Q}_1+\frac{1}{\sigma'}\mathbf{P}\right]\times\mathbf{N}\\
&=n_1\cos r\left[\frac{-1}{\rho'_q}(-\mathbf{N}\sin r+\overline{\mathbf{Q}}\cos r)\times\mathbf{N}-\frac{1}{\sigma'}\overline{\mathbf{Q}}\right]\\
&=-\frac{n_1\cos^2 r}{\rho'_q}\mathbf{P}-\frac{n_1\cos r}{\bar{\sigma}}\overline{\mathbf{Q}}.
\end{aligned} \tag{7.50}$$

$$
\begin{aligned}
(\overline{\mathbf{Q}} \cdot \nabla)\mathbf{S} &= [(\mathbf{S}\sin i + \mathbf{Q}\cos i)\cdot\nabla]\mathbf{S} \\
&= \sin i(\mathbf{S}\cdot\nabla)\mathbf{S} + \cos i(\mathbf{Q}\cdot\nabla)\mathbf{S} \\
&= n\cos i\left[-\frac{1}{\rho_q}\mathbf{Q} + \frac{1}{\sigma}\mathbf{P}\right].
\end{aligned} \tag{7.51}
$$

$$
\begin{aligned}
[(\overline{\mathbf{Q}} \cdot \nabla)\mathbf{S}] \times \mathbf{N} &= n\cos i\left[-\frac{1}{\rho_q}(\mathbf{Q}\times\mathbf{N}) + \frac{1}{\sigma}(\mathbf{P}\times\mathbf{N})\right] \\
&= n\cos i\left[-\frac{1}{\rho_q}(-\mathbf{N}\sin i + \overline{\mathbf{Q}}\cos i)\times\mathbf{N} - \frac{1}{\sigma}\overline{\mathbf{Q}}\right] \\
&= -\frac{n\cos^2 i}{\rho_q}\mathbf{P} - \frac{n\cos i}{\sigma}\overline{\mathbf{Q}}.
\end{aligned} \tag{7.52}
$$

$$
\begin{aligned}
\mathbf{S}_1 \times [(\overline{\mathbf{Q}} \cdot \nabla)\mathbf{N}] &= \mathbf{S}_1 \times \left[-\frac{1}{\bar{\rho}_q}\overline{\mathbf{Q}} + \frac{1}{\bar{\sigma}}\mathbf{Q}_1\right] \\
&= -\frac{1}{\bar{\rho}_q}\mathbf{S}_1 \times (\mathbf{S}_1\sin r + \mathbf{Q}_1\cos r) + \frac{1}{\bar{\sigma}}(\mathbf{S}_1\times\mathbf{P}) \\
&= \frac{n_1\cos r}{\bar{\rho}_q}\mathbf{P} + \frac{n_1}{\bar{\sigma}}\mathbf{Q}_1.
\end{aligned} \tag{7.53}
$$

$$
\begin{aligned}
\mathbf{S} \times [(\overline{\mathbf{Q}} \cdot \nabla)\mathbf{N}] &= \mathbf{S} \times \left[-\frac{1}{\bar{\rho}_q}\overline{\mathbf{Q}} + \frac{1}{\bar{\sigma}}\mathbf{Q}_1\right] \\
&= -\frac{1}{\bar{\rho}_q}\mathbf{S} \times (\mathbf{S}\ \sin i + \mathbf{Q}\cos i) + \frac{1}{\bar{\sigma}}(\mathbf{S}\times\mathbf{P}) \\
&= \frac{n\cos i}{\bar{\rho}_q}\mathbf{P} + \frac{n}{\bar{\sigma}}\mathbf{Q}.
\end{aligned} \tag{7.54}
$$

Substituting Eqs. 7.50, 7.52, 7.53 and 7.54 back into Eq. 7.48 yields

$$
\begin{aligned}
&-\frac{n_1\cos^2 r}{\rho'_q}\mathbf{P} - \frac{n_1\cos r}{\sigma'}\overline{\mathbf{Q}} + \frac{n_1\cos r}{\bar{\rho}_q}\mathbf{P} + \frac{1}{\bar{\sigma}}(n\mathbf{Q} + \gamma\overline{\mathbf{Q}}) \\
&= -\frac{n\cos i}{\rho_q} - \frac{n\cos i}{\sigma}\overline{\mathbf{Q}} + \frac{n\cos i}{\bar{\rho}_q}\mathbf{P} + \frac{n}{\bar{\sigma}}\overline{\mathbf{Q}}
\end{aligned} \tag{7.55}
$$

Collecting terms in Eq. 7.55, we get

$$
\begin{aligned}
&\mathbf{P}\left[-\frac{n_1\cos^2 r}{\rho'_q} + \frac{n_1\cos r}{\bar{\rho}_q} + \frac{n\cos^2 i}{\rho_q} - \frac{n\cos i}{\bar{\rho}_q}\right] \\
&+\overline{\mathbf{Q}}\left[-\frac{n_1\cos r}{\bar{\sigma}} + \frac{\gamma}{\bar{\sigma}} + \frac{n\cos i}{o}\right] = 0.
\end{aligned} \tag{7.56}
$$

As before we set the coefficients equal to zero and get

$$
\begin{aligned}
\frac{n_1 \cos^2 r}{\rho'_q} &= \frac{n \cos^2 i}{\rho_q} + \frac{n_1 \cos r - n \cos i}{\bar{\rho}_q} \\
&= \frac{n \cos^2 i}{\rho_q} + \frac{\gamma}{\bar{\rho}_q} \\
\frac{n_1 \cos r}{\bar{\sigma}} &= \frac{n \cos i}{\sigma} + \frac{\gamma}{\bar{\sigma}}.
\end{aligned}
\tag{7.57}
$$

Collecting the results of Eqs. 7.47 and 7.57 we obtain

$$
\begin{cases}
\dfrac{n_1 \cos^2 r}{\rho'_q} = \dfrac{n \cos^2 i}{\rho_q} + \dfrac{\gamma}{\bar{\rho}_q} \\
\dfrac{n_1 \cos r}{\sigma'} = \dfrac{n \cos i}{\sigma} + \dfrac{\gamma}{\bar{\sigma}} \\
\dfrac{n_1}{\rho'_p} = \dfrac{n}{\rho_p} + \dfrac{\gamma}{\bar{\rho}_p}.
\end{cases}
\tag{7.58}
$$

The first and third of these are the Coddington equations.

These now provide us with $1/\rho'_q$, $1/\sigma'$ and $1/\rho'_p$.

By inverting the equations in Eq. 7.27 and applying the result to the refracted wave front, we get

$$
\tan 2\theta' = \frac{2/\sigma'}{1/\rho'_p - 1/\rho'_q},
\tag{7.59}
$$

and

$$
\begin{cases}
\dfrac{1}{\rho'_\xi} = \dfrac{\cos^2 \theta'}{\rho'_p} + \dfrac{\sin^2 \theta'}{\rho'_q} - \dfrac{\sin 2\theta'}{\sigma'} \\
\dfrac{1}{\rho'_\eta} = \dfrac{\sin^2 \theta'}{\rho'_p} + \dfrac{\cos^2 \theta'}{\rho'_q} + \dfrac{\sin 2\theta'}{\sigma'}
\end{cases}
\tag{7.60}
$$

which gives us the two principal curvatures of the refracted wave front at the point where ray and wavefront intersect.

One of the principal directions, the vector $\mathbf{T}'$, is given by

$$
\mathbf{T}' = \mathbf{P} \cos \theta' + \mathbf{Q}_1 \sin \theta'.
\tag{7.61}
$$

To conclude and recapitulate, ordinary ray tracing consists of two operations, transfer and refraction. Generalized ray tracing consists of transfer, extended to include the calculation of principal radii of curvature, ordinary refraction, a rotation of component vectors and curvatures to positions normal to and lying in the plane of incidence. This is followed by an application of the augmented Coddington equations as given in Eq. 7.58 followed by another rotation to the principal directions as in Eqs. 7.59 and 7.60.

7.9 The Caustic

We have derived Eqs. 7.58 through 7.61 that yield formulas for the two principal curvatures $1/\rho'_\xi$ and $1/\rho'_\eta$ as well as $\mathbf{T}'$, one of the two principal directions; the second comes from the vector product of $\mathbf{T}'$ and $\mathbf{S}_1$. These processes can be iterated from surface to surface through an optical system. Suppose we start at some fixed object point and trace a ray passing through that point through the optical system to image space. The ray trace will terminate at some surface, the last refracting surface or the exit pupil. At that point there passes a wavefront whose principal directions are given by $\mathbf{T}'$ and $\mathbf{T}' \times \mathbf{S}_1$ and whose two principal curvatures are $1/\rho'_\xi$ and $1/\rho'_\eta$.

The reciprocals of the principal curvatures, ρ'_ξ and ρ'_η, are the two principal radii of curvature, the distances to the two principal centers of curvature and therefore are the two points lying on the caustic surface. The coordinates of the caustic points can be calculated by using the transfer equation, Eq. 7.2.

One of the definitions of the caustic surface is that it is the locus of the principal centers of curvature and is therefore a surface of two sheets on each of which lies one of the two principal centers of curvature.

7.10 The Prolate Spheroid

An excellent example of the use of generalized ray tracing is its application to the reflecting prolate spheroid whose equations we encountered in Chapter 4. The point of incidence

$$\mathbf{R} = \varrho(\sin\phi\sin\theta,\ \sin\phi\cos\theta,\ \cos\phi), \tag{4.75}$$

where the radius vector ϱ was derived in Chapter 2 as

$$\varrho = \frac{r}{1-\epsilon\cos\phi}. \tag{2.36}$$

We will also need ϕ', the angle opposite ϕ

$$\begin{cases} \sin\phi' = (1-\epsilon^2)\sin\phi/\mathcal{K}^2, \\ \cos\phi' = \left[2\epsilon-(1+\epsilon^2)\cos\phi\right]/\mathcal{K}^2, \end{cases} \tag{2.38}$$

where

$$\mathcal{K}^2 = (1+\epsilon^2) - 2\epsilon\cos\phi, \tag{2.39}$$

as well as ϱ' the distance between the point on the surface and the distal focus

$$\varrho' = \frac{r\mathcal{K}^2}{(1-\epsilon^2)(1-\epsilon\cos\phi)} = \frac{\varrho\,\mathcal{K}^2}{1-\epsilon^2}. \tag{2.40}$$

There we also obtained the differential element of area

$$D^2 = EG - F^2 = \frac{\varrho^4\sin^2\phi}{(1-\epsilon\cos\phi)^2}\mathcal{K}^2, \tag{4.79}$$

as well as the unit normal vector

$$\mathbf{N} = \frac{1}{\mathcal{K}}\big(\sin\phi\sin\theta,\ \sin\phi\cos\theta,\ \cos\phi{-}\epsilon\big), \tag{4.81}$$

and the principal directions given as unit vectors

$$\begin{cases} \mathbf{T}_+ = \dfrac{1}{\mathcal{K}}\big((\cos\phi - \epsilon)\sin\theta,\ (\cos\theta - \epsilon)\cos\theta,\ -\sin v\big) \\ \mathbf{T}_- = (\cos\theta,\ -\sin\theta,\ 0). \end{cases} \tag{4.84}$$

The principal curvatures are

$$\begin{cases} \dfrac{1}{\rho_+} = -\dfrac{(1 - \epsilon\cos\phi)^2}{\varrho\,\mathcal{K}^3} \\ \dfrac{1}{\rho_-} = -\dfrac{1}{\varrho\,\mathcal{K}}\,. \end{cases} \tag{4.85}$$

7.11 Rays in the Spheroid

Now we come to a situation where notational problems arise and confuse. We are used to dealing with $\mathbf{S} = (u,\ v,\ w)$ where $\mathbf{S}^2 = n^2$, the square if the refractive index. However when dealing with reflective systems the incident ray and its reflected counterpart are, save for sign, equal. In the equation for reflection, Eq. 7.11, the refractive index of the medium cancels out from the two members of the equation leaving a relation involving only direction cosines. So, in this section and in other chapters where reflection is dominant we will use the notation $\mathbf{S} = (\xi,\ \eta,\ \zeta)$ where $\mathbf{S}^2 = \xi^2 + \eta^2 + \zeta^2 = 1$.

Suppose there is a ray that passes through the proximal focus of a spheroid with the direction cosine vector $\mathbf{S} = (\xi,\ \eta,\ \zeta)$ so that, from Eq. 4.75

$$\begin{aligned} &\sin\phi = \sqrt{\xi^2 + \eta^2} = \sqrt{1 - \zeta^2}, && \cos\phi = \zeta, \\ &\sin\theta = \xi/\sqrt{1 - \zeta^2}, && \cos\theta = \eta/\sqrt{1 - \zeta^2}. \end{aligned} \tag{7.62}$$

The point of incidence will be

$$\mathbf{R} = \varrho(\xi,\ \eta,\ \zeta), \tag{7.63}$$

where, from Eqs. 2.36 and 2.40, we have

$$\varrho = \frac{r}{1 - \epsilon\zeta} \qquad \varrho' = \varrho\frac{\mathcal{K}^2}{1 - \epsilon^2} = \frac{r\mathcal{K}^2}{(1 - \epsilon\zeta)(1 - \epsilon^2)}. \tag{7.64}$$

Then the unit normal vector, from Eq. 4.81, is,

$$\mathbf{N} = \frac{1}{\mathcal{K}}(\xi,\ \eta,\ \zeta - \epsilon), \tag{7.65}$$

where, from Eq. 2.39

$$\mathcal{K}^2 = 1 + \epsilon^2 - 2\epsilon\,\zeta. \tag{7.66}$$

The angle opposite ϕ, the angle between a reflected ray and the axis of the spheroid, from Eq. 2.38, is

$$\begin{cases} \sin\phi' = (1-\epsilon^2)\sqrt{1-\zeta^2}/\mathcal{K}^2, \\ \cos\phi' = \left[2\epsilon - (1+\epsilon^2)\zeta\right]/\mathcal{K}^2, \end{cases} \tag{7.67}$$

The calculation of the cosine and sine of the angle of incidence is straightforward

$$\begin{cases} \cos i = \mathbf{S}\cdot\mathbf{N} = \dfrac{1}{\mathcal{K}}(\xi,\,\eta,\,\zeta)\cdot(\xi,\,\eta,\,\zeta-\epsilon) = \dfrac{1-\epsilon\zeta}{\mathcal{K}} \\ \sin i = \dfrac{\epsilon\sqrt{1-\zeta^2}}{\mathcal{K}}, \end{cases} \tag{7.68}$$

so that, from Eq. 7.11

$$\mathbf{S}_1 = -\mathbf{S} + 2\,\frac{1-\epsilon\zeta}{\mathcal{K}}\,\mathbf{N}, \tag{7.69}$$

from which comes the direction vector of the reflected ray

$$\begin{aligned} \mathbf{S}_1 &= \frac{1}{\mathcal{K}}\left((1-\epsilon^2)\xi,\,(1-\epsilon^2)\eta,\,(1+\epsilon^2)\zeta - 2\,\epsilon\right) \\ &= \frac{1}{\mathcal{K}}\left[(1-e^2)\mathbf{S} - 2\epsilon(1-\epsilon\zeta)\mathbf{Z}\right], \end{aligned} \tag{7.70}$$

where $\mathbf{Z}$ is the unit vector along the z-axis.

From Eq. 7.22 we can calculate $\mathbf{P}$, the unit vector normal to the plane of incidence

$$\mathbf{P} = \frac{\mathbf{S}\times\mathbf{N}}{\sin i} = \frac{1}{\sqrt{1-\zeta^2}}(-\eta,\,\xi,\,0), \tag{7.71}$$

and from Eq. 7.23 we find $\mathbf{Q}$, the unit vector lying in the plane of incidence, to be

$$\mathbf{Q} = \mathbf{S}\times\mathbf{P} = \frac{-1}{\sqrt{1-\zeta^2}}(\zeta\,\xi,\,\zeta\,\eta,\,\zeta^2-1) = \frac{-1}{\sqrt{1-\zeta^2}}(\zeta\,\mathbf{S} - \mathbf{Z}). \tag{7.72}$$

Again using Eq. 7.23 we calculate $\overline{\mathbf{Q}}$ the vector that corresponds to $\mathbf{Q}$ but is associated with the reflecting surface rather than with the incident wavefront

$$\overline{\mathbf{Q}} = \mathbf{N}\times\mathbf{P} = \frac{-1}{\mathcal{K}^2\sqrt{1-\zeta^2}}\left(\xi(\zeta-\epsilon),\,\eta(\zeta-\epsilon),\,\zeta^2-1\right), \tag{7.73}$$

and with this and Eq. 7.25 we can find its counterpart after reflection

$$\begin{aligned}\mathbf{Q}_1 &= -\mathbf{Q} + 2\,\frac{1-\epsilon\zeta}{\mathcal{K}}\,\overline{\mathbf{Q}} \\ &= \frac{-1}{\mathcal{K}\sqrt{1-\zeta^2}}\Big(\xi\big[\zeta(1+\epsilon^2)-2\epsilon\big], \\ &\qquad \eta\big[\zeta(1+\epsilon^2)-2\epsilon\big], -(1-\zeta^2)(1-\epsilon^2)\Big).\end{aligned} \tag{7.74}$$

Now return to the reflecting surface, the prolate spheroid. From Eq. 4.84 we get the principal directions

$$\begin{cases}\overline{\mathbf{T}}_+ = \dfrac{1}{\mathcal{K}}\dfrac{1}{\sqrt{1-\zeta^2}}\left((\zeta-\epsilon)\xi,\ (\zeta-\epsilon)\eta,\ -(1-\zeta^2)\right) \\ \overline{\mathbf{T}}_- = \dfrac{1}{\sqrt{1-\zeta^2}}(\eta,\ -\xi,\ 0),\end{cases} \tag{7.75}$$

and from Eq. 4.85, the principal curvatures

$$\begin{cases}\dfrac{1}{\overline{\rho}_+} = -\dfrac{(1-\epsilon\zeta)^2}{\varrho\,\mathcal{K}^3} \\ \dfrac{1}{\overline{\rho}_-} = -\dfrac{1}{\varrho\,\mathcal{K}}.\end{cases} \tag{7.76}$$

The vector $\overline{\mathbf{T}}$ can be in either of the two principal directions. We choose the simplest

$$\overline{\mathbf{T}} = -\mathbf{T}_- = \frac{-1}{\sqrt{1-\zeta^2}}(\eta,\ -\xi,\ 0) = \mathbf{P}, \tag{7.77}$$

this from Eq. 7.75, so that, from Eq. 7.28

$$\cos\bar{\theta} = \overline{\mathbf{T}}\cdot\mathbf{P} = 1. \tag{7.78}$$

Now almost everything is set up for the application of the equations for generalized ray tracing. Lacking only are details of the incident wavefront. Suppose that an object point is located along a chief ray at a distance t in front of the proximal focus of the spheroid as shown in Fig. 7.3. Then the radius of the spherical wavefront at the point of incidence is $\varrho + t$ where ϱ is from Eq. 7.64. The wavefront is a sphere; all of its points are umbilical and the principal directions are not defined. Both principal curvatures are equal to each other and to the reciprocal of the radius of the spherical wavefront at the point of incidence. The vector $\mathbf{T}$ is therefore ambiguous and we can choose it to be anything convenient; we let it equal to $\mathbf{P}$ so that

$$\cos\theta = 1, \tag{7.79}$$

this from Eq. 7.26. From Eqs. 7.78 and 7.79 it follows that $\theta = \bar{\theta} = 0$. Because of this $\sigma = 1/\bar{\sigma} = 0$ from Eqs. 7.27 and 7.29. It is also evident that the equations for rotation are redundant.

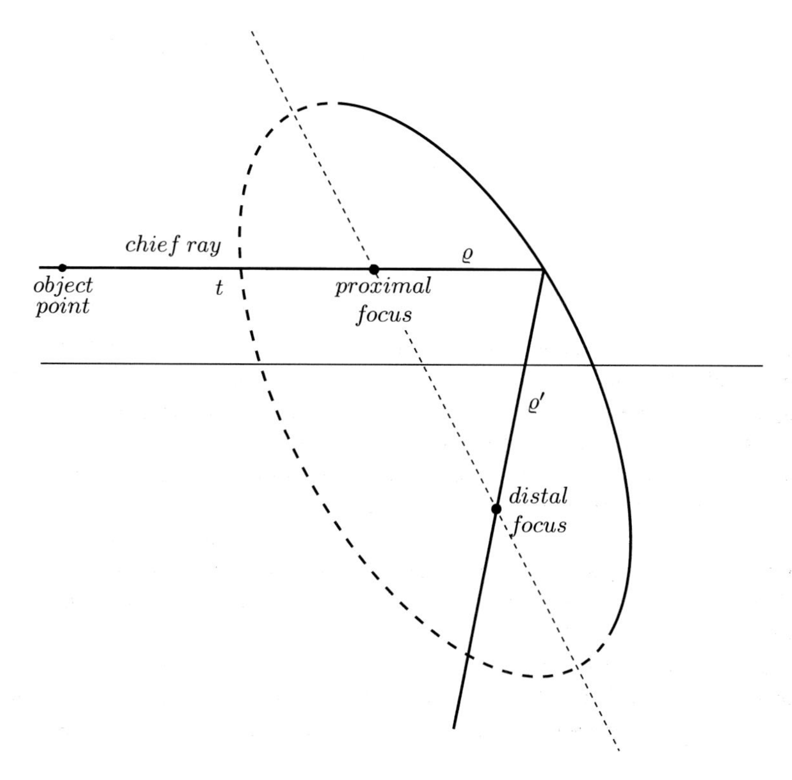

Figure 7.3: *Spheroid with Proximal and Distal Foci.* A ray through the proximal focus.

Now we can invoke the equations for generalized ray tracing found in Eq. 7.58. From this the second equation of Eq. 7.58 comes

$$\frac{1}{\sigma'} = 0. \tag{7.80}$$

The remaining two equations of Eq. 7.58 constitute the original Coddington equations. They are

$$\begin{cases} \dfrac{1}{\rho'_q} = -\dfrac{1}{\rho_q} + \dfrac{2\cos i}{\overline{\rho}_q} \\ \dfrac{1}{\rho'_p} = -\dfrac{1}{\rho_p} + \dfrac{2}{\overline{\rho}_p \cos i}. \end{cases} \tag{7.81}$$

By making substitutions using Eqs. 7.68, 7.76 and 4.85, we get

$$\begin{cases} \dfrac{1}{\rho'_q} = -\dfrac{1}{\rho_q} + \dfrac{2}{r}\dfrac{(1-\epsilon\zeta)^2}{\mathcal{K}} \\ \dfrac{1}{\rho'_p} = -\dfrac{1}{\rho_p} + \dfrac{2}{r}\dfrac{(1-\epsilon\zeta)^2}{\mathcal{K}}. \end{cases} \tag{7.82}$$

In Eq. 7.76 we have identified ρ_+ with ρ_p and ρ_- with ρ_q.

Now if the incident wavefront is a sphere then $\rho_p = \rho_q = \rho$, the sphere's radius. In that case $\rho'_p = \rho'_q$ and the reflected wavefront in the neighborhood of the traced ray has the property that the two principal curvatures are equal. That is not to say that the reflected wavefront is a sphere; only that it has an umbilical point at the traced ray. The radius of curvature of the wavefront at this point, say ρ' is then given by this simpler version of Eq. 7.82

$$\frac{1}{\rho'} = -\frac{1}{\rho} + \frac{2}{r}\frac{(1-\epsilon\zeta)^2}{\mathcal{K}}. \tag{7.83}$$

Suppose the object point is at the proximal focus so that $\rho = \varrho$, where ϱ is given in Eq. 7.64. Then from the above we get

$$\frac{1}{\rho'} = \frac{1-\epsilon\zeta}{r\mathcal{K}}[-\mathcal{K} + 2(1-\epsilon\,\zeta)] = \frac{(1-\epsilon\,\zeta)(1-\epsilon^2)}{r\mathcal{K}} = \frac{1}{\varrho'}, \tag{7.84}$$

where ϱ' is also found in Eq. 7.64. The image point is then at the distal focus, just as it should be.

Now refer to Fig. 7.4 which shows the same ellipse encountered in the earlier illustrations. Assume that $\overline{AB}$ and $\overline{BC}$ are rays incident and reflected with point of incidence B. Suppose that the object point is at A and that its conjugate is at C. Let A be located a distance d in front of the proximal focus and let C be at distance d' in front of the distal focus. In what follows we adopt the usual optical convention that the curvature of a surface is positive if it is concave to the right. Thus a converging wavefront has a positive curvature; if it diverges, its curvature is negative. A wavefront originating at a point A will be spherical and its radius of curvature at the point of incidence, B will be $\varrho + d$; its curvature is negative. We know that the reflected wavefront will have an umbilical point at B and that at that point its radius of curvature will be $\varrho' - d'$. Since it is converging the sign of its curvature will be positive. Then Eq. 7.83 becomes

$$\frac{1}{\varrho' + d'} = -\frac{1}{\varrho + d} + \frac{2}{r}\frac{(1-\epsilon\zeta)^2}{\mathcal{K}}. \tag{7.85}$$

By taking the difference of Eqs. 7.83 and 7.85 we get the very useful formula

$$\frac{1}{\varrho' + d'} - \frac{1}{\varrho'} = -\left(\frac{1}{\rho - d} - \frac{1}{\rho}\right), \tag{7.86}$$

which reduces to, using Eqs. 7.64 and 7.66

$$d' = \frac{dr\mathcal{K}^2}{(1-\epsilon^2)[r(1-\epsilon^2) - 2d(1-\epsilon\zeta)^2]}. \tag{7.87}$$

But what we really want is the reciprocal, the curvature of the wavefront and that is

$$\frac{1}{d'} = -\frac{(1-\epsilon^2)}{\mathcal{K}^2}\left[\frac{1-\epsilon^2}{t} + 2\frac{(1-\epsilon\zeta)^2}{r}\right]. \tag{7.88}$$

These expressions will be seen again in Chapter 11 when they will be applied to the Modern Schiefspiegler.

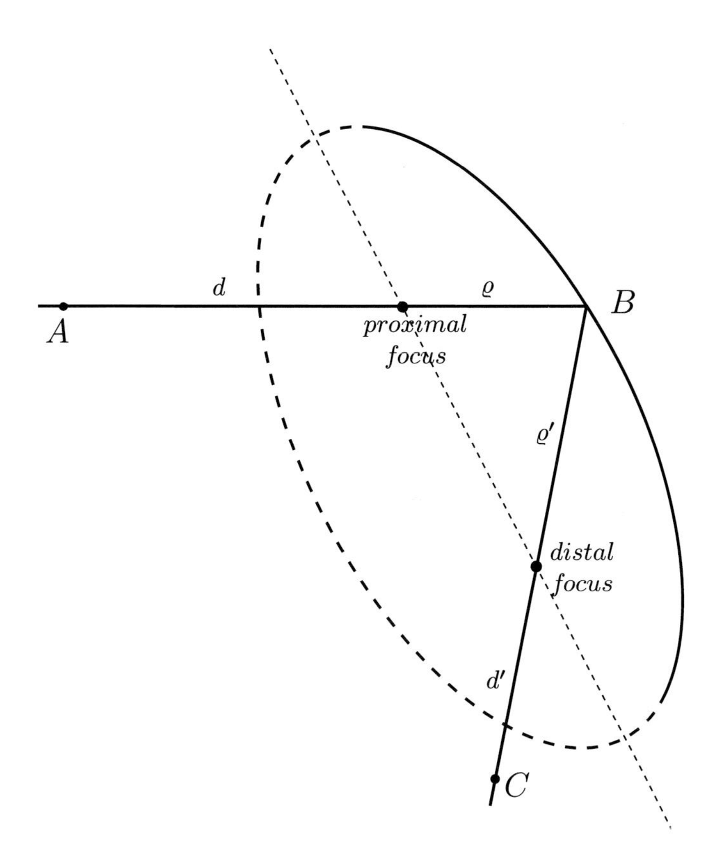

Figure 7.4: *A Ray Through Proximal and Distal Foci.*

Using generalized ray tracing, starting with a fixed object point, the points on the caustic can be calculated and plotted just as are spot diagrams. The resulting three dimensional picture is the aberrated image of the object point. The caustic has an obvious advantage over the spot diagram in that it is three dimensional and is not dependent on the location of a particular plane in image space.

A difficulty is that the caustic is a two-sheeted surface making it hard to plot and difficult to interpret once plotted. A way around this is to plot the two sheets separately. However these equations by themselves can not indicate which sheet the centers of curvature belong to. In plotting a sequence of these point pairs and comparing their coordinates with those of the next preceding pair it is possible to assign these points to the correct surface.

This completes the study of the mathematics of ray tracing and its generalization along with an example of its application to generalized rays in a prolate spheroid. By changing only the eccentricity ϵ these results can be extended to all conic surfaces of revolution. The results obtained here will prove to be useful in subsequent chapters in particular Chapter 11 on the Modern Schiefspiegler.

8 Aberrations in Finite Terms

Recall that the eikonal function is a measure of the optical path length between some arbitrary but fixed object point and a point $\mathbf{P}$ in image space. In the notation we use, the coordinates of the object point are never specified; the arguments of the function are the components of $\mathbf{P}$ in image space. We have already seen that if $\mathbf{P}$ is a perfect image of the object point; every ray that passes through the object point must also passes through $\mathbf{P}$. In that case the eikonal function is constant.

Solutions of the eikonal equation have been obtained previously in the form of a power series. Since perfect imagery requires the eikonal to be constant, all coefficients of the power series, except for the constant term, must be *zero*. It follows that non zero coefficients of this power series must be measures of departures from perfect image formation. These coefficients have come to be identified as *aberrations*; in the past they have proved to be eminently useful. Indeed, the third order Seidel aberrations or primary aberrations have been indispensable in optical design for the better part of a century. In modern optical design these aberration coefficients have been extended to higher degrees.

But this power series has never been shown to converge nor even to be summable. It may even be that the series solution is asymptotic; that the series diverges but that a finite number of terms provide a optimum approximation to real rays as determined by ray tracing. As far as I know this has never been successfully investigated.

In practice the aberration coefficients are calculated surface by surface in a lens design with arguments that are not the coordinates of $\mathbf{P}$ but are heights and slopes of the marginal and principal paraxial rays that are traced through the system. So the result of all this is layers of approximations superimposed on earlier approximations. The miracle is that it not only works but works exceedingly well. But as methods for designing lenses improve and as lenses become better and better the assumptions on which these techniques are based become less valid. The purpose of this chapter is to introduce a novel alternative.

We have already seen that the shape of the caustic surface depends on the k-function and its first and second partial derivatives. We also know that the caustic can represent the monochromatic aberrations that we call *image errors*; *spherical aberration*, *astigmatism* and *coma*. It follows then that all of these image errors are contained in the k-function. The *field errors*, *distortion* and *field curvature*, are determined by the location of the caustic in image space and this also is determined by the k-function. We conclude that all of the monochromatic aberrations reside in and can be determined by the k-function.

In all our discussions we have excluded color entirely as has been explained in the introductory chapter. But we do not exclude the possibility that the k-function can be generalized to

The Mathematics of Geometrical and Physical Optics: The k-function and its Ramifications. O.N. Stavroudis

ISBN: 3-527-40448-1

accommodate a dependence on wavelength and, in that case, provide a model for the chromatic aberrations.

So, we visualize the k-function as a closed form function of the variables v and w that is a representation of the total aberrations associated with an optical system relative to some fixed object point. The coefficients of the commonly used infinite series expansion are identified with various aberrations that had been observed in severely aberrated systems. Hence the title of this chapter, *Aberrations in Finite Terms*.

8.1 Herzberger's Diapoints

In Chapter 7 we mentioned spot diagrams, obtained by intersecting traced rays with a plane in image space, and how it can be used to interpret the aberrations of an optical system. Herzberger proposed an alternative approach applicable only to rotationally symmetric optical systems. The plane determined by an off axis object point and the axis of rotational symmetry is called the *meridional plane*. *Meridional rays* are defined as those rays that lie entirely in the meridional plane at every medium in an optical system. This can be seen from the ray tracing equations for transfer (Eqs. 7.2 and 7.3) and for refraction (7.9 and 7.10). *Skew rays* are defined as those rays that are not meridional; rays that leave the object point at an angle to the meridional plane and pass through the optical system to image space where once again they intersects the meridional plane. (They may also cross the meridional plane in the interior of the lens.) The point of intersection of a skew ray with the meridional plane in image space is Herzberger's *diapoint*.[1]

The totality of rays from a fixed object point that pass through the lens produce a distribution of diapoints on the meridional plane in image space just as they form a distribution of points in a spot diagram. Unlike the spot diagram, their distribution does not depend on the position of an image plane; indeed, diapoints are useful in determining the best location of an image plane. On the other hand the spot diagram does provide a good approximation to the real image of a point object.

8.2 Herzberger's Fundamental Optical Invariant

This is a differential invariant that is valid for all optical systems, including off-axis telescopes and inhomogeneous media.[2] Consider a lens that consists of three surfaces as shown in Fig. 8.1. The vectors $\mathbf{Q}$, $\mathbf{Q}_1$, $\mathbf{Q}_2$, $\mathbf{Q}_3$, and $\mathbf{Q}'$, all referred to some arbitrary coordinate origin, represent points on a ray traced through the system. Moreover $\mathbf{Q}_1$, $\mathbf{Q}_2$, $\mathbf{Q}_3$, lie on the three refracting surfaces; $\mathbf{Q}$ and $\mathbf{Q}'$ are points in object space and image space, respectively that may or may not be conjugates. Let n, n_1, n_2, and n' be the refractive indices of the four media and let $\mathbf{C}$, $\mathbf{C}_1$, $\mathbf{C}_2$ and $\mathbf{C}'$ be the appropriate ray direction cosine vectors and let $\mathbf{N}_1$, $\mathbf{N}_2$, $\mathbf{N}_3$ be the unit normal vectors to the refracting surfaces at $\mathbf{Q}_1$, $\mathbf{Q}_2$, $\mathbf{Q}_3$, respectively. Now let λ, λ_1, λ_2 and λ' be the lengths of the ray segments between the points of incidence. Then, from the

[1]Herzberger 1946, Herzberger 1954.

[2]Herzberger 1935, Stavroudis 1972a, Stavroudis 1972b.

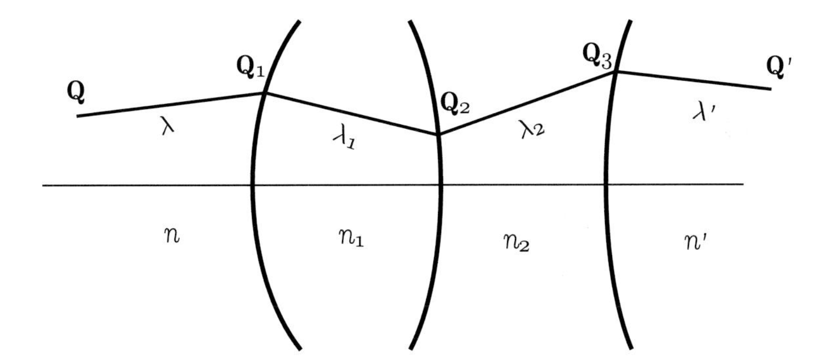

Figure 8.1: *Ray Through a Sample System.* Shown is a ray – $\mathbf{Q},\ \mathbf{Q}_1,\ \mathbf{Q}_2,\ \mathbf{Q}_3,\ \mathbf{Q}'$ – traced through this simple optical system. Refractive indices are $n,\ n_1,\ n_2,$ and n' where these media are separated by the three surfaces. The lengths of the ray segments are $\lambda,\ \lambda_1,\ \lambda_2$ and λ' so that its optical path length is $n\lambda + n_1\lambda_1 + n_2\lambda_2 + n'\lambda'$.

transfer equation from Chapter 7,

$$\overline{\mathbf{P}} = \mathbf{P} + \lambda\mathbf{S}, \tag{7.2}$$

as well as the refraction equation,

$$\mathbf{S}_1 = \mathbf{S} + \gamma\mathbf{N}, \tag{7.9}$$

the following ray tracing equations hold,

$$\begin{cases} \mathbf{Q}_1 = \mathbf{Q} + \lambda\mathbf{C} & \\ \mathbf{Q}_2 = \mathbf{Q}_1 + \lambda_1\mathbf{C}_1, & n_1\mathbf{C}_1 = n\,\mathbf{C} + \gamma_1\mathbf{N}_1 \\ \mathbf{Q}_3 = \mathbf{Q}_2 + \lambda_2\mathbf{C}_2, & n_2\mathbf{C}_2 = n_1\mathbf{C}_1 + \gamma_2\mathbf{N}_2 \\ \mathbf{Q}' = \mathbf{Q}_3 + \lambda'\mathbf{C}', & n'\mathbf{C}' = n_2\mathbf{C}_2 + \gamma_3\mathbf{N}_3. \end{cases} \tag{8.1}$$

Now assume that this ray is a member of a two-parameter family of rays that depends on the parameters p and q and then take the derivatives of the $\mathbf{Q}'s$ with respect to p. The result is,

$$\begin{cases} \dfrac{\partial\mathbf{Q}_1}{\partial p} = \dfrac{\partial\mathbf{Q}}{\partial p} + \lambda\dfrac{\partial\mathbf{C}}{\partial p} + \dfrac{\partial\lambda}{\partial p}\mathbf{C} \\ \dfrac{\partial\mathbf{Q}_2}{\partial p} = \dfrac{\partial\mathbf{Q}_1}{\partial p} + \lambda_1\dfrac{\partial\mathbf{C}_1}{\partial p} + \dfrac{\partial\lambda_1}{\partial p}\mathbf{C}_1 \\ \dfrac{\partial\mathbf{Q}_3}{\partial p} = \dfrac{\partial\mathbf{Q}_2}{\partial p} + \lambda_2\dfrac{\partial\mathbf{C}_2}{\partial p} + \dfrac{\partial\lambda_2}{\partial p}\mathbf{C}_2 \\ \dfrac{\partial\mathbf{Q}'}{\partial p} = \dfrac{\partial\mathbf{Q}_3}{\partial p} + \lambda_3\dfrac{\partial\mathbf{C}_3}{\partial p} + \dfrac{\partial\lambda_3}{\partial p}\mathbf{C}_3 \end{cases} \tag{8.2}$$

Now scalar multiply the first of these by $n\,\mathbf{C}$, the second by $n_1\mathbf{C}_1$, the third by $n_2\mathbf{C}_2$, and the fourth by $n_3\mathbf{C}_3$ and add. Recall that $\mathbf{C}_i^2 = 1$ so that $\partial\mathbf{C}_i/\partial p \cdot \mathbf{C}_i = 0$ so we get,

$$
\begin{aligned}
&n\mathbf{C}\cdot\frac{\partial\mathbf{Q}_1}{\partial p} + n_1\mathbf{C}_1\cdot\frac{\partial\mathbf{Q}_2}{\partial p}, + n_2\mathbf{C}_2\cdot\frac{\partial\mathbf{Q}_3}{\partial p} + n_3\mathbf{C}_3\cdot\frac{\partial\mathbf{Q}'}{\partial p}\\
&= n\mathbf{C}\cdot\frac{\partial\mathbf{Q}}{\partial p} + n_1\mathbf{C}_1\cdot\frac{\partial\mathbf{Q}_1}{\partial p} + n_2\mathbf{C}_2\cdot\frac{\partial\mathbf{Q}_2}{\partial p} + n_3\mathbf{C}_3\cdot\frac{\partial\mathbf{Q}_3}{\partial p}\\
&+ n\frac{\partial\lambda}{\partial p} + n_1\frac{\partial\lambda_1}{\partial p} + n_2\frac{\partial\lambda_2}{\partial p} + n_3\frac{\partial\lambda_3}{\partial p}.
\end{aligned}
\tag{8.3}
$$

These are rearranged to get,

$$
\begin{aligned}
&n\mathbf{C}\cdot\frac{\partial\mathbf{Q}}{\partial p} + \frac{\partial\mathbf{Q}_1}{\partial p}\cdot(n_1\mathbf{C}_1 - n\mathbf{C}) + \frac{\partial\mathbf{Q}_2}{\partial p}\cdot(n_2\mathbf{C}_2 - n_1\mathbf{C}_1)\\
&+\frac{\partial\mathbf{Q}_3}{\partial p}\cdot(n_3\mathbf{C}_3 - n_2\mathbf{C}_2) - n_3 \quad \mathbf{C}_3\cdot\frac{\partial\mathbf{Q}'}{\partial p} + \frac{\partial\mathcal{L}}{\partial p} = 0.
\end{aligned}
\tag{8.4}
$$

where $\mathcal{L}$ is the optical path length, $\mathcal{L} = n\lambda + n_1\lambda_1 + n_2\lambda_2 + n_3\lambda_3$.

But from Eq. 8.1 $n_i\mathbf{C}_i - n_{i-1}\mathbf{C}_{i-1} = \gamma_i\mathbf{N}_i$ so that each term in Eq. 8.4 containing these quantities is multiplied by $\partial\mathbf{Q}_i/\partial p$. This is the derivative of a point on the refracting surface with respect to the arbitrary parameter and therefore is a vector tangent to the surface. The scalar product of this with the normal vector $\mathbf{N}_i$ is zero so all of the terms in Eq. 8.4 vanish except for three, resulting in,

$$
n\mathbf{C}\cdot\frac{\partial\mathbf{Q}}{\partial p} - n_3\mathbf{C}_3\cdot\frac{\partial\mathbf{Q}'}{\partial p} + \frac{\partial\mathcal{L}}{\partial p} = 0.
\tag{8.5}
$$

With respect to the second parameter q the exact same steps yield a second equation,

$$
n\mathbf{C}\cdot\frac{\partial\mathbf{Q}}{\partial q} - n_3\mathbf{C}_3\cdot\frac{\partial\mathbf{Q}'}{\partial q} + \frac{\partial\mathcal{L}}{\partial q} = 0.
\tag{8.6}
$$

We assume that second derivatives exist and are continuous so that if we take the derivative of Eq. 8.5 with respect to p and the derivative of Eq. 8.6 with respect to q and then eliminate $\partial^2\mathcal{L}/\partial u\partial v$ between them we get,

$$
n'\left(\frac{\partial\mathbf{Q}'}{\partial p}\cdot\frac{\partial\mathbf{C}'}{\partial q} - \frac{\partial\mathbf{Q}'}{\partial q}\cdot\frac{\partial\mathbf{C}'}{\partial p}\right) = n\left(\frac{\partial\mathbf{Q}}{\partial p}\cdot\frac{\partial\mathbf{C}}{\partial q} - \frac{\partial\mathbf{Q}}{\partial q}\cdot\frac{\partial\mathbf{C}}{\partial p}\right).
\tag{8.7}
$$

The quantity,

$$
n\left(\frac{\partial\mathbf{Q}}{\partial p}\cdot\frac{\partial\mathbf{C}}{\partial q} - \frac{\partial\mathbf{Q}}{\partial q}\cdot\frac{\partial\mathbf{C}}{\partial v}\right)
\tag{8.8}
$$

has not only the same value in object and image space but also in every intermediate space in the optical system. This is the *Fundamental Optical Invariant.*

8.3 The Lens Equation

Now lets let the points $\mathbf{Q}$ and $\mathbf{Q}'$ lie on planes in object and image space that pass through the coordinate origin so that $z = z' = 0$. Set $n\mathbf{C} = \mathbf{S} = (u,\ v,\ w)$, $n'\mathbf{C}' = \mathbf{S}' = (u',\ v',\ w')$ making $\mathbf{S}$ and $\mathbf{S}'$ reduced direction cosine vectors. We have that $\mathbf{Q} = (x,\ y,\ ,0)$ and $\mathbf{Q}' = (x',\ y',\ ,0)$ so that Eq. 8.7 becomes,

$$x'_p u'_q + y'_p v'_q - x'_q u'_p - y'_q v'_p = x_p u_q + y_p v_q - x_q u_p - y_q v_p. \tag{8.9}$$

This Herzberger called the *direct method*.[3]

The parameters p and q are arbitrary; they may take on any values we choose. We select for these parameters all possible combinations of the object space variables, $x,\ y,\ u,\ v$ resulting in a system of six equations,

$$\begin{cases} x'_x u'_y + y'_x v'_y - x'_y u'_x - y'_y v'_x = 0 \\ x'_x u'_u + y'_x v'_u - x'_u u'_x - y'_u v'_x = 1 \\ x'_x u'_v + y'_x v'_v - x'_v u'_x - y'_v v'_x = 0 \\ x'_y u'_u + y'_y v'_u - x'_u u'_y - y'_u v'_y = 0 \\ x'_y u'_v + y'_y v'_v - x'_v u'_y - y'_v v'_y = 1 \\ x'_u u'_v + y'_u v'_v - x'_v u'_u - y'_v v'_u = 0. \end{cases} \tag{8.10}$$

This can be expressed in Matrix form. Let $\mathbf{M}$ be given by,

$$\mathbf{M} = \begin{pmatrix} x'_x & x'_y & x'_u & x'_v \\ y'_x & y'_y & y'_u & y'_v \\ u'_x & u'_y & u'_u & u'_v \\ v'_x & v'_y & v'_u & v'_v, \end{pmatrix} \tag{8.11}$$

and let $\mathbf{J}$ be the constant matrix,

$$\mathbf{J} = \begin{pmatrix} 0 & 0 & 1 & 0 \\ 0 & 0 & 0 & 1 \\ -1 & 0 & 0 & 0 \\ 0 & -1 & 0 & 0 \end{pmatrix}. \tag{8.12}$$

Then Eq. 8.10 can be written as,

$$\mathbf{M}\,\mathbf{J}\,\mathbf{M}^{\mathbf{T}} = \mathbf{J}, \tag{8.13}$$

where $\mathbf{M}^{\mathbf{T}}$ is the transpose of $\mathbf{M}$. This I have called the *lens equation*.[4] It is clear that $|\mathbf{M}|$ is the Jacobian of this transformation.

[3] Herzberger 1943a, Herzberger 1943b.
[4] Stavroudis 1972a, pp. 245–249.

We can think of a lens as a device for mapping rays in object space into rays in image space. Recall that a ray is determined by a point on a reference plane, $(x,\ y,\ ,0)$, and a direction, $(u,\ v,\ w)$. We can think of this transformation as the system of equations,

$$\begin{cases} x' = x'(x,\ y,\ u,\ v) \\ y' = y'(x,\ y,\ u,\ v) \\ u' = u'(x,\ y,\ u,\ v) \\ v' = v'(x,\ y,\ u,\ v). \end{cases} \tag{8.14}$$

Since conjoining two lenses results in yet another lens this system of transforms can be regarded as a *Lie group*[5], in which the identity element is simply the lack of a lens and the inverse transform is obtained by solving the equations in Eq. 8.14 for $(x,\ y,\ u,\ v)$ in terms of $(x',\ y',\ u',\ v')$.

We can go even further. Equation 8.13 is the defining equation for one of the two classical matrix groups, the *symplectic group* (the other being the *orthogonal group*).[6] From Eq. 8.13 we can show that $\mathbf{M}$ is a unitary matrix. Note first of all that $|\,\mathbf{J}\,| = 1$ so that by taking the determinant of both members of Eq. 8.13 we get $|\,\mathbf{M}\,|^2 = 1$. It follows that $\mathbf{M}$ is non singular; inverses always exist.

We remarked earlier that a new lens was obtained where two lenses were conjoined. This is accomplished by multiplying the corresponding two matrices. This works because the product of two Jacobian matrices is yet another Jacobian.

Suppose we decompose a lens into its elementary components, a sequence of alternating operations, refraction and transfer, each capable of being represented by a $\mathbf{M}$ matrix so that the lens could be represented by a matrix product. The idea was to apply this to the design an analysis of lenses; like all such approaches to optical design it was far to complicated to be useful.[7]

8.4 Aberrations in Finite Terms

This work is an extension and a continuation of work first published over twenty years ago.[8] It is based on the general solution of the eikonal equation and the k-function, the expression for the wavefront train defined by that solution

$$\mathbf{W}(v,\ w;\ s) = \frac{q}{n^2}\mathbf{S} - \mathbf{K}, \tag{5.72}$$

where

$$q = (ns{-}k){-}(vk_v + wk_w) = (ns{-}k){+}\mathbf{S}{\cdot}\mathbf{K}. \tag{5.71}$$

[5] Stavroudis 1972a, pp. 285–286.
[6] Stavroudis 1972a, pp. 286–288.
[7] Stavroudis 1959.
[8] Stavroudis and Fronczek 1978a.

and the caustic surface associated with that train,

$$\mathbf{C}_{\pm}(v,\, w) = \frac{1}{2n^2}(H \pm \mathcal{S})\mathbf{S} - \mathbf{K}, \tag{6.21}$$

where,

$$H = (n^2 - v^2)k_{vv} + (n^2 - w^2)k_{ww} - 2vwk_{vw}, \tag{6.6}$$

where T^2 is the Hessian determinant,

$$T^2 = k_{vv}k_{ww} - k_{vw}^2, \tag{6.7}$$

and where $\mathcal{S}$ is the discriminant of a quadratic polynomial,

$$\mathcal{S}^2 = H^2 - 4n^2u^2T^2. \tag{6.15}$$

The parameter s in Eq. 5.71 is a distance measured from some fixed but arbitrary starting point.

We choose this starting point to be one particular wave front in the train, which I've called the *archetype*; with its optical path length from the ultimate object point equal to *zero*. This is of little importance in this chapter but will be useful in Chapter 9 which involves the refraction of the k-function. We are concerned with the structure of the caustic surface in which resides all of the properties, the aberrations if you will, of the wavefront train. These are uniquely and completely determined by the k-function.

The sign ambiguity, as we have already seen, shows that the caustic is a two sheeted surface. Equation 5.71 shows that there is a plane of symmetry. In what follows we will assume that we are dealing with an optical system possessing rotational symmetry and that the plane of symmetry is the meridional plane determined by some object point.

Herzberger used the distribution of diapoint to define a hierarchical system of aberrations.[9] In general, the diapoints cover a region of the meridional plane. This he called *deformation error*. If the diapoints collapse to an arc they form the *half-symmetric* image. If the arc degenerates to a straight line the image is *symmetric*. Finally, if the straight line degenerates to a single point the image is *sharp*.

8.5 Half-Symmetric, Symmetric and Sharp Images

Now recall the vector $\mathbf{K} = (0,\, k_v,\, k_w)$ a vector function of v and w that lies on the meridianal plane. Every ray in the system defined by Eq. 5.72 is associated with a particular pair of values, v and w, that define a value of the vector $\mathbf{K}$ that, in turn, defines a point on the meridional plane. Therefore $\mathbf{K}$ defines the aggregate of diapoints associated with the system.

With no restrictions on $\mathbf{K}(v,\, w)$ it represents deformation error. We will impose restrictions on $\mathbf{C}_{\pm}$, given in Eq. 6.21, and observe their effect on $\mathbf{K}$.

We first set T^2, defined in Eq. 6.7, equal to zero; T^2 is the Hessian determinant of the k-function. It can also be regarded as a Jacobian determinant of the two functions k_v and k_w and its vanishing implies that a functional relation exists between them, say,

$$k_w = f(k_v). \tag{8.15}$$

[9] Herzberger 1950.

Then,

$$\begin{cases} k_{vw} = f'\, k_{vv} \\ k_{ww} = f'^2\, k_{vv}, \end{cases} \tag{8.16}$$

so that, from Eq. 6.6,

$$H = k_{vv}[(n^2 - v^2) - 2vwf' + (n^2 - w^2)f'^2], \tag{8.17}$$

and, from Eq. 6.21,

$$\begin{cases} \mathbf{C}_+ = -\mathbf{K} + \frac{1}{n^2} H\, \mathbf{S} \\ \mathbf{C}_- = -\mathbf{K}. \end{cases} \tag{8.18}$$

In this case the negative branch of the caustic surface, $\mathbf{C}_-$, degenerates into an arc on the meridional plane and is therefore the half-symmetric image.

Now consider Eq. 8.16 as a non linear first order partial differential equation. With the method of Lagrange and Charpit that we derived in Chapter 5, we can get a complete integral and a general solution. First from Eq. 8.16 we define,

$$\mathcal{F} = f(k_v) - k_w = f(p) - q = 0, \tag{8.19}$$

that leads us to,

$$\mathcal{F}_v = \mathcal{F}_w = \mathcal{F}_k = 0, \quad \mathcal{F}_p = f', \quad \mathcal{F}_q = -1.$$

As before we introduce the symbols p and q to represent the two partial derivatives. When this is plugged into the characteristic equations Eq. 5.29 we get

$$\frac{dp}{0} = \frac{dq}{0} = \frac{-dv}{f'} = \frac{-dw}{-1} = \frac{-dk}{pf' - q}.$$

The simplest of these and the one we choose to work with is $dp = 0$ which leads us to the characteristic $p = \alpha$. This leads us to $q = f(\alpha)$, from Eq. 8.19, which we use in the total differential equation,

$$dk = \alpha dv + f(\alpha)dw, \tag{8.20}$$

which after integration yields the complete integral,

$$k(v,\, w) = \alpha v + f(\alpha)w + \beta(\alpha), \tag{8.21}$$

where $\alpha = \alpha(v,\, w)$. By appending to this the side condition,

$$v + f'(\alpha)\, w\, + \beta'(\alpha) = 0, \tag{8.22}$$

we get the general solution. It follows that $k_v = \alpha$ and $k_w = f(\alpha)$, just as it should be.

From this we can see that $\mathbf{K} = \big(0,\ \alpha,\ f(\alpha)\big)$ is the degenerate caustic and at the same time the half-symmetric image. We take α to be the its parameter each value of which determines one point on the half-symmetric image; at the same time each value of α determines a one-parameter family of rays that pass through that point.

The unit tangent vector to the half symmetric is obtained by differentiating $\mathbf{K}$ with respect to α. This we normalize to get the unit tangent vector

$$\mathbf{T}(\alpha) = \frac{\big(0,\ 1,\ f'(\alpha)\big)}{\sqrt{1 + f'^2}}. \tag{8.23}$$

When α is held fixed $\mathbf{K}$ is a point on the arc and $\mathbf{T}$ is its unit tangent vector at that point. Consider the totality of rays through that point; that is to say, for all those values of v and w for which $\alpha(v,\ w) = constant$. The reduced direction cosine vector of these rays is, of course $\mathbf{S}$. The cosine of the angle, ϕ between one of these rays and the unit tangent vector $\mathbf{T}$ of the degenerate caustic is given by,

$$\cos\phi = \frac{1}{n}(\mathbf{S}\cdot\mathbf{T}) = \frac{v + f'(\alpha)\,w}{n\sqrt{1 + f'^2}} = -\frac{\beta'}{n\sqrt{1 + f'^2}}, \tag{8.24}$$

from Eq. 8.23. But β' and f' are both functions of α which is fixed. Therefore, $\cos\phi$ is constant for all rays that pass through a single point on the half-symmetric image; they must then make the same angle with its tangent and form a cone whose vertex is at that point. This is Herzberger's theorem.[10]

If f and β are linear in α then the arc in Eq. 8.22 becomes a straight line and the image is symmetric. If α is constant the image is sharp; the rays defined by $\alpha(v,\ w) = constant$ all must pass through the sharp image.

Now lets take a closer look at the sharp image. Refer now to Eq. 8.22 and note that, $k = vk_v + w\kappa_w + \beta$. When we substitute this into Eq. 5.72, we get $q = ns - \beta$ and

$$\mathbf{W}(v,\ w) = \frac{1}{n^2}(ns - \beta)\mathbf{S} - \big(0,\ \alpha,\ F(\alpha)\big), \tag{8.25}$$

a train of concentric spherical wavefronts centered on the sharp image, exactly as it should. The sharp image occurs when $s = \beta/n$. This established a connection between Herzberger's system of hierarchial aberrations and the k-function.

[10] Herzberger 1943b.

9 Refracting the k-Function

In Chapter 5 we derived a general solution of the eikonal equation, in vector form that represents a wavefront train in a homogeneous, isotropic medium. This general solution is *almost* universal; *almost* in that it cannot represent a train of plane wavefronts. In that case we must uses the intermediate solution, the *complete integral* as did Luneburg.[1] An immediate consequence of this solution is the vector equation for the caustic surface associated with this train of wavefronts.

These expressions for wavefront and caustic depend in a natural way on two independent parameters, v and w, two of the three reduced ray direction cosines of the orthotomic system of rays associated with the wavefront train; the third is $u = \sqrt{n^2 - v^2 - w^2}$. Another parameter is s, the geometric distance of a wavefront from some arbitrary but fixed starting point. Throughout all of this work this starting point is assumed to be the *archetypical wavefront*, defined as that unique wave front in the train whose optical path length from the ultimate object point is *zero*. More will be said on this subject subsequently.

The immediate problem is to determine the effect a spherical refracting surface has on the k-function. In this chapter we will find the k-function, both for a plane wavefront train that is refracted by a spherical refracting surface and a train of spherical wavefronts refracted by a refracting plane. Then we will deal with more general problem, the refraction of a train of spherical wavefronts, originating at a finite object point, by a sphere.

An important thing to keep in mind is that the caustic surface, its shape and its location, tells us everything about the geometric aberrations associated with an optical system relative to some fixed object point. Indeed, in Chapter 8 we have shown that components of the expression for the caustic can be identified with Herzberger's diapoints.[2]

We now return to the general solution of the eikonal equation and the expression for the wavefront train in a homogeneous, isotropic medium

$$\mathbf{W}(v,\, w,\, s) = \frac{q}{n^2}\mathbf{S} - \mathbf{K}, \tag{5.72}$$

where

$$q = ns - k + \left(v\frac{\partial k}{\partial v} + w\frac{\partial k}{\partial w}\right) = ns - k + \mathbf{S}\cdot\mathbf{K}. \tag{5.71}$$

[1] Stavroudis 1972a, Chapter VIII.
[2] Herzberger 1958, Chapters VII and XXVII.

The Mathematics of Geometrical and Physical Optics: The k-function and its Ramifications. O.N. Stavroudis

ISBN: 3-527-40448-1

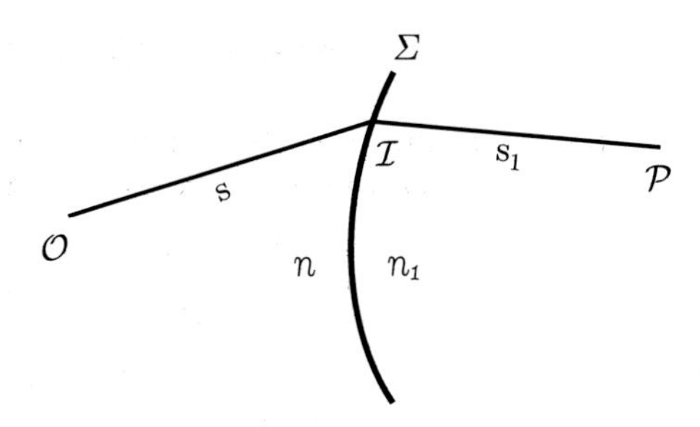

Figure 9.1: *Refraction of Wavefronts.* Refracting surface: Σ. Object point: $\mathcal{O}$. Point of Incidence: $\mathcal{I}$. Point on ray after refraction: $\mathcal{P}$.

These vectors are

$$\begin{cases} \mathbf{S} = (u,\ v,\ w), \qquad \mathbf{S}^2 = n^2 \\ \mathbf{K} = \left(0,\ \dfrac{\partial k}{\partial v},\ \dfrac{\partial k}{\partial w}\right). \end{cases} \tag{5.70}$$

Recall that $\mathbf{S}$ is the reduced ray direction vector and that n is the refractive index of the medium.

The caustic is the locus of the two principal centers of curvature of any wavefront in the train and is given by,

$$\mathbf{C}_{\pm}(v,\ w) = \frac{1}{2n^2}(H \pm \mathcal{S})\mathbf{S} - \mathbf{K}, \tag{6.21}$$

where,

$$H = (n^2 - v^2)\frac{\partial^2 k}{\partial v^2} + (n^2 - w^2)\frac{\partial^2 k}{\partial w^2} - 2vw\frac{\partial^2 k}{\partial v \partial w}, \tag{6.6}$$

$$T^2 = \frac{\partial^2 k}{\partial v^2}\frac{\partial k^2}{\partial w^2} - \left(\frac{\partial k^2}{\partial v \partial w}\right)^2, \tag{6.7}$$

and

$$\mathcal{S} = \sqrt{H^2 - 4n^2u^2T^2}. \tag{6.15}$$

The sign ambiguity in Eq. 6.21 indicates that the caustic is indeed a two sheeted surface.

Consider an optical system that consists of a sequence of spherical refracting surfaces, with centers on a common axis, that separate media of constant refractive index. Choose some arbitrary but fixed point in object space. It is the center of a train of concentric spherical wavefronts incident on the first refracting sphere. Its own k-function is determined by the location of this object point. Indeed the object point itself can be regarded either as the degenerate caustic for this incident wavefront train or as the degenerate 'first' wavefront in the train, a sphere of radius zero.

This train of spherical waves is incident on the refracting surface Σ and, after refraction, gives rise to a wavefront train whose properties are completely determined by its k-function.

Let the object point be $\mathcal{O}$ and let $\mathcal{I}$ be the point of incidence on Σ and let $\mathcal{O}\,\mathcal{I}$ be a ray as shown in Fig. 9.1. Let the geometric distance between $\mathcal{O}$ and $\mathcal{I}$ be s. Then ns is the optical path length to $\mathcal{I}$ where n is the refractive index of the first medium. The ray is refracted at $\mathcal{I}$ as it enters the second medium where the refractive index is n_1. Take any point $\mathcal{P}$ on this refracted ray and let s_1 be the geometric distance from $\mathcal{I}$ to $\mathcal{P}$. Then the optical path length along this refracted ray must be $n_1\, s_1$ where n_1 is the refractive index of the second medium. It follows that the total optical path length from object point $\mathcal{O}$ to $\mathcal{P}$ is $n\, s + n_1\, s_1$.

Earlier we defined the archetype wavefront as that unique wavefront in image space whose optical path measured from the object point is *zero*.[3] Its location is therefore the geometric distance,

$$s_1 = -sn/n_1, \tag{9.1}$$

from the refracting surface. Since s_1 is negative the archetype is to the left of the refracting sphere and is therefore virtual.

This wavefront train is then incident on the second spherical refracting surface giving rise to yet another refracted train in the next succeeding medium. This process is repeated until the image space is reached after the final refracting surface.

Now refer back to the expression for the general wavefront in Eq. 5.72 in which the parameter s appears. In each medium of the optical system this parameter will be measured from the archetype[4] in that medium. All conjugate wavefronts have the same optical path length, $n\, s$, measured from the appropriate archetype which leads us to,

$$ns = n_1 s_1 = n_2 s_2 = \dots\ . \tag{9.2}$$

9.1 Refraction

In what follows we will be concerned with refraction at a single spherical refracting surface.[5] To keep things simple the incident wavefront and its parameters will lack subscripts; the refracted wavefront and its parameters will be signaled by the subscript ($_1$). In what follows r is the radius of the refracting sphere; the coordinate origin is taken at its center. We will write the refracted wavefront train as $\mathbf{W}_1(v_1,\ w_1,\ s_1)$ in the manner of Eq. 5.72 and q_1, $\mathbf{S}_1$ and $\mathbf{K}_1$ defined as in Eqs. 5.70 and 5.71.

It is obvious that any point lying on a refracting surface is its own image; every ray incident at such a point must, after refraction, also pass through that point. It is clear that the optical path *difference* between that point and its image must be *zero*.

A wave front will intersect a refracting sphere in an arc. Choose any point on that curve; there the ray is normal to the wave front and is refracted by the sphere. Through that same point, the refracted ray is normal to the refracted wave front.

The optical path distance from any point on the archetype, along a ray, to its point of intersection with a wave front, is constant. Refer now to the point on this refracting surface, through which two wave fronts pass, one before refraction and one after. The optical path

[3] Stavroudis and Fronczek 1976.

[4] Stavroudis 1995a.

[5] Stavroudis 1995b.

difference between this point before refraction and its image after refraction was shown to be *zero*. It follows that the optical path distance from any point on the incident wavefront, along an incident ray and its refracted counterpart, to the refracted wave front, must also equal *zero*. We have shown that any two wave fronts that pass through a point on a refracting surface must be conjugates and therefore, from Eq. 9.1, $s_1 = -sn/n_1$. This result we will use later

Now refer back to Chapter 7, on ray tracing. Recall

$$\mathbf{S}_1 = \mathbf{S} + \gamma\mathbf{N}, \tag{7.9}$$

and

$$\gamma = \mathbf{S}_1 \cdot \mathbf{N} - \mathbf{S} \cdot \mathbf{N} = n_1 \cos r - n \cos i, \tag{7.10}$$

where *i* is the angle of incidence and where *r* is the angle of refraction. The two vectors, **S** and $\mathbf{S}_1$ are referred to as *reduced direction vectors*. By writing Eq. 7.9 as $\gamma\mathbf{N} = \mathbf{S}_1 - \mathbf{S}$ and squaring both sides we get

$$\gamma^2 = (\mathbf{S}_1 - \mathbf{S})^2 = n_1^2 + n^2 - 2\alpha, \tag{9.3}$$

where

$$\alpha = \mathbf{S}_1 \cdot \mathbf{S} = uu_1 + vv_1 + ww_1. \tag{9.4}$$

9.2 The Refracting Surface

Our goal is to obtain $k_1(v_1, w_1)$ from $k(v, w)$ when the latter is known.[6] To do so we need two vector functions to represent the refracting sphere, one in terms of v and w associated with the incident wavefront train, the other in terms of v_1 and w_1 the parameters for the refracted wavefronts. The first of these involves $k(v, w)$ and its derivatives; the second, $k_1(v_1, w_1)$. Since both are representations of the same surface we are able to equate the two and obtain the required relationships.

The equation of a sphere centered at the coordinate origin is,

$$\mathcal{F} = \mathbf{P}^2 - r^2 = 0. \tag{9.5}$$

Its gradient is normal to the surface so that the unit normal vector **N** is then,

$$\mathbf{N} = \frac{\nabla\mathcal{F}}{\sqrt{(\nabla\mathcal{F})^2}} = \frac{1}{r}\mathbf{P}. \tag{9.6}$$

We next substitute the expression for the incident wavefront train, Eq. 5.74, into the equation for the sphere, Eq. 9.5 and get a quadratic equation in q,

$$q^2 - 2aq + n^2(b^2 - r^2) = 0, \tag{9.7}$$

[6] Stavroudis 1969.

where

$$\begin{cases} a \;= v\dfrac{\partial k}{\partial v} + w\dfrac{\partial k}{\partial w} \\ b^2 = \left(\dfrac{\partial k}{\partial v}\right)^2 + \left(\dfrac{\partial k}{\partial w}\right)^2. \end{cases} \tag{9.8}$$

The solution of Eq. 9.7 is,

$$q = a - \Delta, \tag{9.9}$$

where

$$\Delta^2 = a^2 - n^2(b^2 - r^2). \tag{9.10}$$

In Eq. 9.8 we defined a and b. By comparing Eqs. 9.9 and 5.71 we can see that

$$k - ns = \Delta. \tag{9.11}$$

We have chosen the negative branch of the quadratic solution to place the point of incidence nearer to the vertex of the refracting surface. By substituting the value of q from Eq. 9.9 into Eq. 5.72 we get the equation for the refracting surface in terms of the parameters of the incident ray,

$$\mathbf{P}(v,\, w) = \frac{1}{n^2}(a - \Delta)\mathbf{S} - \mathbf{K}. \tag{9.12}$$

We go through exactly the same steps for the refracted wavefront train to get,

$$q_1^2 - 2a_1 q_1 + n_1^2(b_1^2 - r^2) = 0, \tag{9.13}$$

an analog of Eq. 9.7, whose solution is (we again choose the negative branch), referring to Eq. 5.71,

$$q_1 = n_1 s_1 - k_1 + a_1 = a_1 - \Delta_1, \tag{9.14}$$

where,

$$\begin{cases} a_1 \;= v_1\dfrac{\partial k_1}{\partial v_1} + w_1\dfrac{\partial k_1}{\partial w_1} \\ b_1^2 \;= \left(\dfrac{\partial k_1}{\partial v_1}\right)^2 + \left(\dfrac{\partial k_1}{\partial w_1}\right)^2 \\ \Delta_1^2 = a_1^2 - n_1^2(b_1^2 - r^2). \end{cases} \tag{9.15}$$

We also get the analog of Eq. 9.12

$$\mathbf{P}_1(v_1,\, w_1) = \frac{1}{n_1^2}(a_1 - \Delta_1)\mathbf{S}_1 - \mathbf{K}_1 \tag{9.16}$$

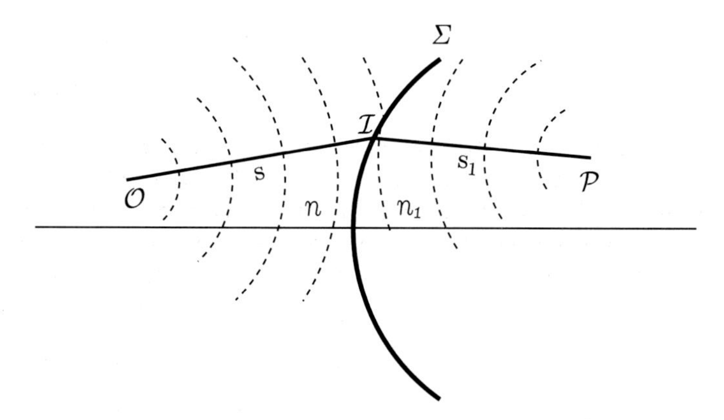

Figure 9.2: *Conjugate Wavefronts.* Refracting sphere: Σ. Object point: $\mathcal{O}$. Point of Incidence: $\mathcal{I}$. Incident wavefront and refracted wavefront are shown.

as well as of Eq. 9.11

$$k_1 - n_1 s_1 = \Delta_1. \tag{9.17}$$

Figure 9.2 shows the refracting sphere Σ and a ray incident at a point $\mathcal{I}$ where it is refracted. One member of the wavefront train incident on the refracting sphere will pass through $\mathcal{I}$; one member of the refracted wavefront train will also pass through this point so that these two wavefronts are conjugates and therefore, as we have already seen, $n\,s = n_1\,s_1$. Applying this to Eqs. 9.11 and 9.17,

$$k_1 = k + \Delta_1 - \Delta. \tag{9.18}$$

Now we find three equations relating the incident k-function with its refracted counterpart. The unit normal vector to the refracting surface must be the same in terms of both the incident and refracted parameters so using Eqs. 9.12 and 9.16 we get,

$$\mathbf{N} = \frac{1}{n^2 r}(a - \Delta)\mathbf{S} - \frac{1}{r}\mathbf{K} = \frac{1}{n_1^2 r}(a_1 - \Delta_1)\mathbf{S}_1 - \frac{1}{r}\mathbf{K}_1. \tag{9.19}$$

In scalar form this is the three equations,

$$\begin{cases} \dfrac{u}{n^2}(a - \Delta) & = \dfrac{u_1}{n_1^2}(a_1 - \Delta_1) \\ \dfrac{v}{n^2}(a - \Delta) - \dfrac{\partial k}{\partial v} & = \dfrac{v_1}{n_1^2}(a_1 - \Delta_1) - \dfrac{\partial k_1}{\partial v_1} \\ \dfrac{w}{n^2}(a - \Delta) - \dfrac{\partial k}{\partial w} & = \dfrac{w_1}{n_1^2}(a_1 - \Delta_1) - \dfrac{\partial k_1}{\partial w_1}. \end{cases} \tag{9.20}$$

In the next step we get three more equations relating the two k-functions. Refer back to Eq. 9.19 to which we apply the vector form of Snell's law, $\mathbf{S} \times \mathbf{N} = \mathbf{S}_1 \times \mathbf{N}$, which leads directly to,

$$\mathbf{S} \times \mathbf{K} = \mathbf{S}_1 \times \mathbf{K}_1, \tag{9.21}$$

which in scalar form is,

$$\begin{cases} w_1 \dfrac{\partial k_1}{\partial v_1} - v_1 \dfrac{\partial k_1}{\partial w_1} = w \dfrac{\partial k}{\partial v} - v \dfrac{\partial k}{\partial w} \\ u_1 \dfrac{\partial k_1}{\partial v_1} = u \dfrac{\partial k}{\partial v} \\ u_1 \dfrac{\partial k_1}{\partial w_1} = u \dfrac{\partial k}{\partial w}. \end{cases} \tag{9.22}$$

These seven equations, Eqs. 9.18, 9.20 and 9.22 are precisely the equations that relate the k-function of the incident wavefront train to that of the refracted wavefront train.

9.3 The Partial Derivatives

By eliminating $\partial k/\partial v$ and $\partial k/\partial w$ between the three equation of Eq. 9.22 we obtain

$$(u_1 w - u w_1)\frac{\partial k_1}{\partial v_1} - (u_1 v - u v_1)\frac{\partial k_1}{\partial w_1} = 0, \tag{9.23}$$

which tells us that there must be some factor A such that

$$\frac{\partial k_1}{\partial v_1} = A(u_1 v - u v_1), \qquad \frac{\partial k_1}{\partial w_1} = A(u_1 w - u w_1). \tag{9.24}$$

This in turn leads us to, using Eq. 9.22

$$\frac{\partial k}{\partial v} = \frac{u_1}{u} A(u_1 v - u v_1), \qquad \frac{\partial k}{\partial w} = \frac{u_1}{u} A(u_1 w - u w_1). \tag{9.25}$$

Now if we let

$$A = B/u_1 \tag{9.26}$$

these partial derivatives take a more symmetric form

$$\frac{\partial k}{\partial v} = \frac{B}{u}(u_1 v - u v_1), \qquad \frac{\partial k}{\partial w} = \frac{B}{u}(u_1 w - u w_1) \tag{9.27}$$

and

$$\frac{\partial k_1}{\partial v_1} = \frac{B}{u_1}(u_1 v - u v_1), \qquad \frac{\partial k_1}{\partial w_1} = \frac{B}{u_1}(u_1 w - u w_1). \tag{9.28}$$

From the first equation of Eq. 9.20 we get

$$\frac{a - \Delta}{n^2} = \frac{u_1}{u}\frac{a_1 - \Delta_1}{n_1^2}, \tag{9.29}$$

which, when substituted into the second equation in Eq. 9.20 yields

$$\begin{aligned}
v\frac{u_1}{u}\frac{a_1-\Delta_1}{n_1^2} - \frac{B}{u}(u_1v-uv_1) &= v_1\frac{a_1-\Delta_1}{n_1^2} - \frac{B}{u_1}(u_1v-uv_1) \\
\frac{a_1-\Delta_1}{n_1^2}(u_1v-uv_1) &= B(u_1v-uv_1)\frac{u_1-u}{u_1} \\
\frac{a_1-\Delta_1}{n_1^2} &= B\frac{u_1-u}{u_1},
\end{aligned} \tag{9.30}$$

where Eqs. 9.24 and 9.25 have been used. From this and Eq. 9.26 it also follows that

$$\frac{a-\Delta}{n^2} = B\frac{u_1-u}{u}. \tag{9.31}$$

The third equation of 9.20 yields identical results. Note that by eliminating $B(u_1-u)$ between Eqs. 9.30 and 9.31 we get back the first equation of Eq. 9.20.

Next we apply the results of Eq. 9.31 to Eqs. 9.8 and 9.10 to get

$$\begin{cases}
a &= \dfrac{B}{u}(n^2u_1-\alpha u) \\
b^2 &= \dfrac{B^2}{u^2}(n^2u_1^2+n_1^2u^2-2\alpha uu_1) \\
\Delta^2 &= B^2(\alpha^2-n^2n_1^2)+r^2n^2
\end{cases} \tag{9.32}$$

and from Eqs. 9.15 and 9.27 the result is

$$\begin{cases}
a_1 &= \dfrac{B}{u_1}(\alpha u_1-n_1^2u) \\
b_1^2 &= \dfrac{B^2}{u_1^2}(n^2u_1^2+n_1^2u^2-2\alpha uu_1) \\
\Delta_1^2 &= B^2(\alpha^2-n^2n_1^2)+r^2n_1^2,
\end{cases} \tag{9.33}$$

where α is defined in Eq. 9.4.

Now go back to Eq. 9.31 and rewrite it as

$$u\Delta = ua - n^2B(u_1-u), \tag{9.34}$$

to which we apply the first equation of Eq. 9.32 to get

$$\Delta = B(n^2-\alpha). \tag{9.35}$$

Now we square both sides and use the third equation of Eq. 9.32. The result is

$$\Delta^2 = B^2(\alpha^2-n^2n_1^2)+r^2n^2 = B^2(n^2-\alpha)^2. \tag{9.36}$$

This we solve for B

$$B^2 = \frac{r^2 n^2}{(n^2 - \alpha)^2 - (\alpha^2 - n^2 n_1^2)}, \tag{9.37}$$

which reduces to

$$B^2(n^2 + n_1^2 - 2\alpha) = B^2\gamma^2 = r^2, \tag{9.38}$$

where we have used γ defined in Eq. 9.3. Our result is an expression for B

$$B = r/\gamma, \tag{9.39}$$

where we have chosen the positive branch of the square root.

We substitute this value of B into Eq. 9.27

$$\begin{cases} \dfrac{\partial k}{\partial v} = \dfrac{r}{u\gamma}(v_1 u - v u_1) \\ \dfrac{\partial k}{\partial w} = \dfrac{r}{u\gamma}(w_1 u - w u_1), \end{cases} \tag{9.40}$$

and again into Eq. 9.28

$$\begin{cases} \dfrac{\partial k_1}{\partial v_1} = \dfrac{r}{u_1\gamma}(v_1 u - v u_1) \\ \dfrac{\partial k_1}{\partial w_1} = \dfrac{r}{u_1\gamma}(w_1 u - w u_1), \end{cases} \tag{9.41}$$

as well as Eq. 9.32

$$\begin{cases} a = \dfrac{r}{u\gamma}(\alpha u - n^2 u_1) \\ b^2 = \dfrac{r^2}{u^2\gamma^2}(n_1^2 u^2 + n^2 u_1^2 - 2\alpha u u_1) \\ \Delta = \dfrac{-r}{\gamma}(n^2 - \alpha), \end{cases} \tag{9.42}$$

and Eq. 9.33

$$\begin{cases} a_1 = \dfrac{r}{u_1\gamma}(n_1^2 u - \alpha u_1) \\ b_1^2 = \dfrac{r^2}{u_1^2\gamma^2}(n_1^2 u^2 + n^2 u_1^2 - 2\alpha u u_1) \\ \Delta_1 = \dfrac{r}{\gamma}(n_1^2 - \alpha). \end{cases} \tag{9.43}$$

In both Eqs. 9.42 and 9.43 square roots were taken to obtain Δ and Δ_1. The signs were chosen in such a way as to be consistent with Eq. 9.18 which, after the appropriate substitutions, becomes

$$\begin{aligned} k_1 &= k + \Delta_1 - \Delta \\ &= k + \frac{r}{\gamma}[(n_1^2 - \alpha) + (n^2 - \alpha)] \\ &= k + \frac{r}{\gamma}(n_1^2 + n^2 - 2\alpha) \\ &= k + r\gamma. \end{aligned} \tag{9.44}$$

Again we use the equation for γ in Eq. 9.3.

9.4 The Finite Object Point

As we did in Chapter 6 we choose an object point with the coordinates $\big(0, -h, -(t+r)\big)$, the starting point of the wavefront train, so that $s = 0$ as shown in Fig. 9.3. It can also be thought of as the caustic associated with this wavefront train. This point can also be thought of as a degenerate spherical wavefront of radius *zero*. Now write Eq. 5.69 in scalar form equating the components equal to the coordinates of the object point. This results in,

$$\begin{cases} x = (q/n^2)u = 0 \\ y = (q/n^2)v - \dfrac{\partial k}{\partial v} = -h \\ z = (q/n^2)w - \dfrac{\partial k}{\partial w} = -(t+r). \end{cases} \tag{9.45}$$

From the first of these it must be that, from Eq. 5.72, recalling that $s = 0$

$$q = -k + \left(v\frac{\partial k}{\partial v} + w\frac{\partial k}{\partial w}\right) = 0, \tag{9.46}$$

and from the second and third equations we can see that,

$$\frac{\partial k}{\partial v} = h, \qquad \frac{\partial k}{\partial w} = t + r. \tag{9.47}$$

When this is applied to Eq. 9.46 we get,

$$k = vh + w(t+r). \tag{9.48}$$

We next apply the relations in Eq. 9.47 to Eq. 9.40 and get,

$$\begin{cases} r(v_1 u - v u_1) = u\gamma h \\ r(w_1 u - w u_1) = u\gamma(t+r). \end{cases} \tag{9.49}$$

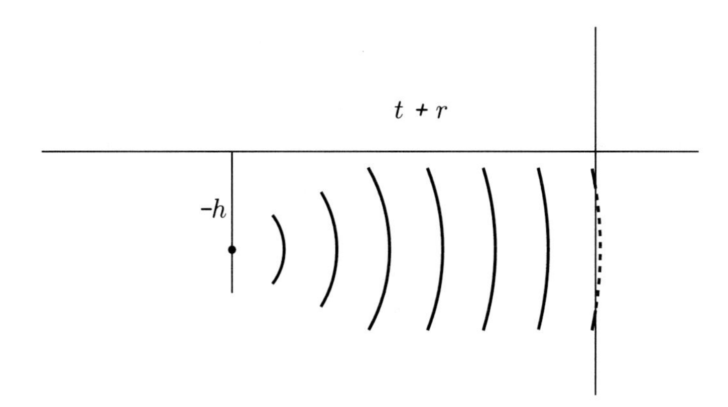

Figure 9.3: *Wavefront Train.* A train of spherical wavefronts centered at $\big(0, -h, -(t+r)\big)$.

Now multiply the first of these by $t+r$ and the second by h and subtract. The result is

$$(v_1u - u_1v)(t+r) - (w_1u - wu_1)h = 0, \tag{9.50}$$

which we solve for u to get

$$u = u_1\frac{v(t+r) - wh}{v_1(t+r) - w_1h}. \tag{9.51}$$

This we substitute back into the first equation of Eq. 9.49. The result is, after some reduction

$$r(vw_1 - v_1w) = \big[v(t+r) - wh\big]\gamma,$$

which we rearrange into

$$v\big[rw_1 - \gamma(t+r)\big] - w(rv_1 - \gamma h) = 0. \tag{9.52}$$

This tells us that there must exist some factor C so that

$$\begin{cases} v = C[rv_1 - \gamma h] \\ w = C[rw_1 - \gamma(t+r)]. \end{cases} \tag{9.53}$$

By substituting this back into Eq. 9.51, the expression for u, we obtain

$$u = Cru_1. \tag{9.54}$$

With these values of the reduced ray direction cosines in object space we can verify that the equations in Eq. 9.47 are satisfied by using Eq. 9.40 and that from Eq. 9.41 we can get

$$\begin{cases} \dfrac{\partial k_1}{\partial v_1} = Crh \\ \dfrac{\partial k_1}{\partial w_1} = Cr(t+r), \end{cases} \tag{9.55}$$

which tells us in no uncertain terms that C must be found.

9.5 The Quest for C

First we calculate $u^2 + v^2 + w^2$ using Eqs. 9.53 and 9.54 to get

$$n^2 = C^2\{r^2 n_1^2 - 2r\gamma[hv_1 + w_1(t+r)] + \gamma^2[h^2 + (t+r)^2]\}$$

which we rearrange into a quadratic polynomial in γ

$$\gamma^2 C^2[h^2 + (t+r)^2] - 2\gamma r C^2[hv_1 + w_1(t+r)] + C^2 r^2 n_1^2 - n^2 = 0. \tag{9.56}$$

Next comes α from Eq. 9.4

$$\alpha = C\{rn_1^2 - \gamma[v_1 h + w_1(t+r)]\}, \tag{9.57}$$

which we plug into the expression for γ^2 from Eq. 9.3 resulting in

$$\gamma^2 = n_1^2 + n^2 - 2C\{rn_1^2 - \gamma[v_1 h + w_1(t+r)]\}$$

which we rearrange into another quadratic polynomial in γ

$$\gamma^2 - 2C\gamma[v_1 h + w_1(t+r)] - [n_1^2 + n^2 - 2Crn_1^2] = 0. \tag{9.58}$$

Between these two polynomials we eliminate γ yielding an equation involving C alone.

To do this we use Sylvester's dialectic method[7] described also in the Appendix. But first to simplify matters we make the following definitions:

$$\begin{cases} M = v_1 h + w_1(t+r) \\ N = h^2 + (t+r)^2, \end{cases} \tag{9.59}$$

so that the two polynomials, Eqs. 9.56 and 9.58 become

$$\begin{cases} \gamma^2 C^2 N - 2\gamma r C^2 M + C^2 r^2 n_1^2 - n^2 = 0 \\ \gamma^2 - 2\gamma CM - (n_1^2 + n^2 - 2Crn_1^2) = 0. \end{cases} \tag{9.60}$$

Sylvester's eliminant between the two equations of Eq. 9.60 is the following determinant equation

$$\begin{aligned} &\begin{vmatrix} C^2N & -2rC^2M & C^2r^2n_1^2 - n^2 & 0 \\ 1 & -2CM & -(n_1^2 + n^2 - 2Crn_1^2) & 0 \\ 0 & C^2N & -2rC^2M & C^2r^2n_1^2 - n^2 \\ 0 & 1 & -2CM & -(n_1^2 + n^2 - 2Crn_1^2) \end{vmatrix} \\ &= \begin{vmatrix} C^2N & -2rC^2M \\ 1 & -2CM \end{vmatrix} \begin{vmatrix} -2C^2rM & C^2r^2n_1^2 - n^2 \\ -2CM & -(n^2 + n_1^2 - 2Crn_1^2) \end{vmatrix} \\ &\quad - \begin{vmatrix} C^2N & C^2r^2n_1^2 - n^2 \\ 1 & -(n^2 + n_1^2 - 2Crn_1^2) \end{vmatrix}^2 = 0. \end{aligned} \tag{9.61}$$

[7] See, for example, Itô 1993, Vol. 1, pp. 369E.

After a long sequence of calculations this reduces to the sextic polynomial equation

$$\begin{aligned} &4n_1^2Nr^2a_1C^6 - 4ra_1a_3C^5 + (4M^2a_4 - a_3^2)C^4 \\ &\quad - 4n^2ra_1C^3 + 2n^2a_3C^2 - n^4 = 0, \end{aligned} \tag{9.62}$$

where

$$\begin{cases} a_1 = M^2 - n_1^2N \\ a_2 = N + r^2 \\ a_3 = n^2N + n_1^2a_2 \\ a_4 = n_1^2r^2 + n^2a_2. \end{cases} \tag{9.63}$$

For convenience let,

$$m_1 = v_1h + w_1(t + r). \tag{9.64}$$

When this and Eqs. 9.53, 9.54 and 9.8 are substituted into Eq. 9.48 we find that,

$$k = \left(\frac{u}{u_1}\right)\left(m_1 - \frac{\gamma}{r}b^2\right), \tag{9.65}$$

which, when applied to Eq. 9.44, yields,

$$k_1 = \left(\frac{u}{u_1}\right)m_1 + \frac{\gamma}{r}\left[r^2 - \left(\frac{u}{u_1}\right)b^2\right]. \tag{9.66}$$

Next square each of the equations in Eqs. 9.53 and 9.54 and add to get,

$$n^2 = \left(\frac{u}{u_1}\right)^2\left[n_1^2 - 2m_1\frac{\gamma}{r} + \left(\frac{\gamma}{r}\right)^2b^2\right], \tag{9.67}$$

where we have used $u^2 + v^2 + w^2 = n^2$ and $u_1^2 + v_1^2 + w_1^2 = n_1^2$.

Next substitute Eq. 9.54 into Eq. 9.55 we find that,

$$\frac{\partial k_1}{\partial v_1} = \left(\frac{u}{u_1}\right)h\,, \qquad \frac{\partial k_1}{\partial w_1} = \left(\frac{u}{u_1}\right)(t + r). \tag{9.68}$$

By eliminating u/u_1 from the two equations in Eq. 9.68 we get the rather remarkable result,

$$(t + r)\frac{\partial k_1}{\partial v_1} - h\frac{\partial k_1}{\partial w_1} = 0, \tag{9.69}$$

a linear, first order, partial differential equation with constant coefficients. By using the method of Lagrange we find its general solution to be an arbitrary function of its characteristic m_1, defined in Eq. 9.64. The derivatives of $k_1(m_1)$ are then,

$$\frac{\partial k_1}{\partial v_1} = k_1'h, \qquad \frac{\partial k_1}{\partial w_1} = k_1'(t + r), \tag{9.70}$$

where k_1' denotes the derivative of k_1 with respect to m_1. By comparing the derivatives in Eq. 9.68 with those in Eq. 9.70 we see that,

$$k_1' = u/u_1. \tag{9.71}$$

When we apply these to Eq. 9.15 we get,

$$a_1 = k_1' m_1, \tag{9.72}$$

and,

$$b_1^2 = k_1'^2 b^2. \tag{9.73}$$

Next refer back to Eq. 9.58 into which we substitute from Eqs. 9.55, 9.64, 9.70 and 9.77, below, to get the quadratic polynomial equation,

$$r\gamma^2 - 2m_1 k_1' \gamma - rM = 0. \tag{9.74}$$

A second quadratic polynomial equation results when the same substitution along with Eqs. 9.8 and 9.47 is made into Eq. 9.56,

$$b^2 k_1'^2 \gamma^2 - 2rm_1 k_1'^2 \gamma + r^2(n_1^2 k_1'^2 - n^2) = 0, \tag{9.75}$$

Now refer back to Eqs. 9.43 and 9.57 in which we substitute from Eq. 9.73 to get,

$$b_1^2 = k_1'^2 b^2 = \frac{r^2}{\gamma^2}[k_1' + (n_1^2 k_1' - n^2)(k_1' - 1)],$$

or,

$$k_1'(k_1' b^2 \gamma^2 - r^2) = r^2(n_1^2 k_1' - n^2)(k_1' - 1). \tag{9.76}$$

In exactly the same way Eq. 9.59 becomes,

$$M = n^2 + n_1^2 - 2\, n_1^2 k_1', \tag{9.77}$$

and Eq. 9.66 turns into,

$$r(k_1 - k_1' m_1) = \gamma(r^2 - k_1' b^2). \tag{9.78}$$

9.6 Developing the Solution

The first step is to eliminate γ from the various polynomial equations given above using Eqs. 9.76 and 9.78 which take the form,

$$\gamma = r\,\frac{m_1\, k_1' - k_1}{b^2\, k_1' - r^2}, \qquad \gamma^2 = r^2\,\frac{(n_1^2\, k_1' - n^2)(k_1' - 1)}{k_1'\,(b^2\, k_1' - r^2)}. \tag{9.79}$$

When we eliminate γ between the two equations in Eq. 9.79 we get,

$$(b^2 k_1' - r^2)(n_1^2 k_1' - n^2)(k_1' - 1) = k_1'(m_1 k_1' - k_1)^2. \tag{9.80}$$

Next substitute the expressions in Eq. 9.79 into the polynomial equation in Eq. 9.76 to get,

$$r^2(n_1^2 k_1' - n^2)(k_1' - 1) - 2m_1 k_1'^2(m_1 k_1' - k_1) - k_1'(b^2 k_1' - r^2)M = 0; \tag{9.81}$$

the same substitution into Eq. 9.75 results in,

$$\begin{aligned} &b^2 k_1'(n_1^2 k_1' - n^2)(k_1' - 1) - 2m_1 k_1'^2(m_1 k_1' - k_1) \\ &+ (b^2 k_1' - r^2)(n_1^2 k_1'^2 - n^2) = 0. \end{aligned} \tag{9.82}$$

Now let,

$$L = (n_1^2 k_1' - n^2)(k_1' - 1). \tag{9.83}$$

Then,

$$\frac{dL}{dk_1'} = n_1(k_1' - 1) + n_1' k_1' - n^2 = -M. \tag{9.84}$$

When these are substituted into Eq. 9.81 we get,

$$r^2 L - 2m_1 k_1'^2(m_1 k_1' - k_1) + k_1'(b^2 k_1'^2 - r^2)dL/dk_1' = 0,$$

which turns into,

$$\frac{d}{d\,k_1'}\left[\frac{L}{k_1'}(b^2 k_1' - r^2)\right] - 2m_1(m_1 k_1' - k_1) = 0. \tag{9.85}$$

Now note that,

$$k_1''\frac{d}{d\,k_1'} = \frac{d}{d\,m_1} \quad \text{and} \quad \frac{d}{d\,m_1}(m_1 k_1' - k_1)^2 = 2k_1'(m_1 k_1' - k_1),$$

so that k_1'' is an integrating factor for Eq. 9.85,

$$\frac{d}{d\,m_1}\left[\frac{L}{k_1'}(b^2 k_1' - r^2) - (m_1 k_1' - k_1)^2\right] = 0.$$

A quadrature results in,

$$(b^2 k_1' - r^2)(n_1^2 k_1' - n^2)(k_1' - 1) = k_1'(m_1 k_1' - k_1)^2 + A^2 k_1' \tag{9.86}$$

where A is a constant of integration. Here we have used Eq. 9.83.

9.7 Conclusions

In this chapter we have attempted to determine the k-function of a refracted wavefront train, $k_1(v_1, w_1)$ from that of an incident wavefront train, $k(v, w)$. A number of abstract relations between these two functions have been found but exact relations are not obtainable without specific knowledge of the incident k-function. We had simplified the calculations by placing the coordinate origin at the center of the refracting sphere.

The idea of the conjugate wavefronts was also introduced. Its definition is reflective, symmetric and transitive and so is an equivalence relation[8] that partitions the wavefront train in an optical system, that arises from some fixed object point, into disjoint equivalence classes, each of which containing exactly one member of the wavefront train in each of the media. In this context the archetypical wavefronts all belong to the same equivalence class.

The ultimate goal of this effort is to develop a system of optical design. The work presented here and in previous papers is, by no means complete. The obvious next step is to obtain solutions for k_1 for specific objects, namely, the finite object point and the object point at infinity. These refracted wavefront trains will then be incident on the next refracting surface presenting yet another problem to be solved using the general methods described here.

That only spherical refracting surfaces were considerd may seem to be an unnecessary restriction, a restriction that excludes much of modern optics. Ultimately we plan to extend these results to at least conic surfaces as well as to systems that are not rotationally symmetric.

[8] Ibid. pp. 530–531.

10 Maxwell Equations and the k-Function

About a century ago Hertz solved the Maxwell equations for a train of spherical wavefronts and in doing so he introduced the concept of the vector and scalar potential functions.[1] He did this by first converting the arguments of the Maxwell equations from Cartesian to spherical coordinates, obtained a wave equation for the potential functions and found a solution that separates in a natural way into a near field solution that anticipated the dipole oscillator, and a far field solution that forms the basis of modern antenna theory.

In Chapter 5 we developed a formula for a wavefront train in a homogeneous isotropic medium as a general solution of the eikonal equation, expressed as a vector function of position. In Chapter 6 we investigated the geometry of these wavefronts and obtained the equation of the caustic surface.

To recapitulate: By holding s fixed and allowing v and w to vary we generate a surface that is one of the wavefronts in the train. The wavefront train is then a one-parameter family of these surfaces with s as its parameter. On the other hand if we hold v and w fixed, the expression becomes the equation of a point on the ray, normal to each of the wavefronts in the train, with s as its variable. From this point of view the totality of rays associated with the wavefront train is a two-parameter family of rays with v and w as the parameters. In Chapter 3 we have referred to such a system as an *orthotomic system* or as a *normal congruence*.

The obvious question is whether Hertz's method can be repeated for this more general case. The first step is to define a system of generalized coordinates based on the solution of the eikonal equation. To do this we assume that through every point $(x,\ y,\ z)$ in space (or at least that part of space that interests us) there passes exactly one wavefront, thus fixing the value of the parameter s. This assumption may fail in the neighborhood of a caustic raising a problem that will be discussed later. The location of the point on the wavefront is determined by two other parameters, v and w. This enables us to treat s, v and w as generalized coordinates of a point in space subject to certain reservations in the neighborhood of the caustic.

Now we are able to express gradient, divergence and curl in terms of these new coordinates. The obvious approach is to follow in Hertz's footsteps, to derive the vector and scalar potential functions, but this leads us to a second order wave equation which, unlike the one obtained by Hertz, can best be described as intractable.

A direct approach is more suitable, leading to expressions for $\mathbf{E}$, the electric field vector, and $\mathbf{H}$, the magnetic field vector. These are manageable and depend on the k-function as well as an additional arbitrary vector function $\mathbf{V}$ that is subject to several conditions, as we will show.

[1] Joos 1941, pp. 322ff. Most modern references start with the dipole oscillator and deduce the spherical wavefronts, such as Marion and Heald 1980, pp. 232–245.

The Mathematics of Geometrical and Physical Optics: The k-function and its Ramifications. O.N. Stavroudis

ISBN: 3-527-40448-1

10.1 The Wavefront

From the general solution of the eikonal equation we obtained the vector function for the wavefront train

$$\mathbf{W}(v,\,w,\,s)=\frac{q}{n^2}\mathbf{S}-\mathbf{K}, \tag{5.72}$$

where

$$\begin{cases}\mathbf{S}\;=(u,\,v,\,w)\\ \mathbf{K}=(0,\,k_v,\,k_w),\end{cases} \tag{5.70}$$

and where the scalar factor is,

$$q=ns-k+(vk_v+wk_w)=ns-k+\mathbf{S}\cdot\mathbf{K}. \tag{5.71}$$

For each s, this vector function of two parameters represents a surface in space and is subject to analysis by the general methods of differential geometry as described in Chapter 5. This surface is a member of the one-parameter train of wavefronts whose parameter is s.

The derivatives of $\mathbf{W}$ are the vectors $\mathbf{W}_v$ and $\mathbf{W}_w$ tangent to the wavefront and $\mathbf{W}_s$ as its normal so that $\mathbf{W}_v\cdot\mathbf{W}_s=\mathbf{W}_w\cdot\mathbf{W}_s=0$; $\mathbf{W}_s$ being a unit normal vector of the wave front and is therefore a ray direction cosine vector. Note also that,

$$\mathbf{W}_s=\frac{1}{D}(\mathbf{W}_v\times\mathbf{W}_w)=\frac{1}{n}\mathbf{S}. \tag{4.7}$$

where,

$$D^2=(\mathbf{W}_v\times\mathbf{W}_w)^2. \tag{4.8}$$

10.2 The Maxwell Equations

We begin with the well-known Maxwell equations for homogeneous, isotropic media,[2]

$$\begin{cases}\nabla\times\mathbf{E}=-\dfrac{\mu}{c}\dfrac{\partial\mathbf{H}}{\partial t} & \nabla\cdot\mathbf{E}=0\\[2ex] \nabla\times\mathbf{H}=\;\;\dfrac{\epsilon}{c}\dfrac{\partial\mathbf{E}}{\partial t} & \nabla\cdot\mathbf{H}=0\end{cases} \tag{10.1}$$

where ϵ and μ are constant and are related to the refractive index by $n=\sqrt{\epsilon\mu}$.

Since we will be dealing with derivatives with respect to s, the distance parameter, rather than the time parameter t, a change of variables from t to s is needed. This is done with

$$\frac{\partial}{\partial t}=\frac{ds}{dt}\frac{\partial}{\partial s}=v\frac{\partial}{\partial s}=\frac{c}{n}\frac{\partial}{\partial s}, \tag{10.2}$$

[2]Stavroudis 1972, Chapters VIII and IX. A general solution using the method of Lagrange and Charpit; a detailed account can be found in Forsyth 1996, Chapter IX. See also Stavroudis 1995.

where v is the velocity of light in this medium. The two Maxwell equations that involve the curl operator are changed into

$$\begin{cases} \nabla \times \mathbf{E} = -\dfrac{\mu}{c}\dfrac{\partial \mathbf{H}}{\partial t} = -\sqrt{\dfrac{\mu}{\epsilon}}\dfrac{\partial \mathbf{H}}{\partial s} \\ \nabla \times \mathbf{H} = \dfrac{\epsilon}{c}\dfrac{\partial \mathbf{E}}{\partial t} = \sqrt{\dfrac{\epsilon}{\mu}}\dfrac{\partial \mathbf{E}}{\partial s}. \end{cases} \tag{10.3}$$

10.3 Generalized Coordinates and the Nabla Operator

We have already discussed the generalized coordinates, v, w and s, induced by the solution of the eikonal equation. The nabla operator can then be expressed in terms of these generalized coordinates and therefore so can the Maxwell equations.

We now return to the caveat mentioned earlier. In the first place the transformation from cartesian coordinates to generalized coordinates is not always single valued; through each point in space there may pass more than one wavefront belonging to the train. As we have already stated, this is certainly true in the neighborhood of a caustic, a cusp locus of the wavefront train.[3] However we are concerned with not only the point in space but also with only one of the wavefronts that passes through it. This appears to remove the mathematical ambiguity. Diffractive interference between wavefronts passing through that point is certainly likely but at this point we can do no more than speculate on its nature.

From here on we will refer to the generalized coordinates v, w, and s. Now assume that some function f depends on these, so that,

$$f_v = \nabla f \cdot \mathbf{W}_v, \qquad f_w = \nabla f \cdot \mathbf{W}_w, \qquad f_s = \nabla f \cdot \mathbf{W}_s. \tag{10.4}$$

Multiply the second of these by $\mathbf{W}_v$ the first, by $\mathbf{W}_w$ and subtract to get,

$$\begin{aligned} f_w \mathbf{W}_v - f_v \mathbf{W}_w &= (\nabla f \cdot \mathbf{W}_w)\mathbf{W}_v - (\nabla f \cdot \mathbf{W}_v)\mathbf{W}_w \\ &= \nabla f \times (\mathbf{W}_v \times \mathbf{W}_w) = D(\nabla f \times \mathbf{W}_s). \end{aligned} \tag{10.5}$$

By taking the vector product of this expression with $\mathbf{W}_s$ and using the third equation in Eq. 10.4 we get,

$$\nabla f = \frac{1}{D}\Big[(\mathbf{W}_w \times \mathbf{W}_s) f_v - (\mathbf{W}_v \times \mathbf{W}_s) f_w\Big] + \mathbf{W}_s f_s \tag{10.6}$$

where D is defined in Eq. 4.8.

It can also be shown that divergence and curl in terms of these generalized coordinates are

$$\nabla \cdot \mathbf{V} = \frac{1}{D}\Big[(\mathbf{W}_w \times \mathbf{W}_s) \cdot \mathbf{V}_v - (\mathbf{W}_v \times \mathbf{W}_s) \cdot \mathbf{V}_w\Big] + \mathbf{W}_s \cdot \mathbf{V}_s. \tag{10.7}$$

and

$$\nabla \times \mathbf{V} = \frac{1}{D}\Big[(\mathbf{W}_w \times \mathbf{W}_s) \times \mathbf{V}_v - (\mathbf{W}_v \times \mathbf{W}_s) \times \mathbf{V}_w\Big] + \mathbf{W}_s \times \mathbf{V}_s. \tag{10.8}$$

[3] Stavroudis et al. 1978.

We also have the directional derivative

$$(\mathbf{P}\cdot\nabla)\mathbf{A} = \frac{1}{D}\Big[(\mathbf{W}_s \times \mathbf{P})\cdot\mathbf{W}_w\,\mathbf{A}_v - (\mathbf{W}_s \times \mathbf{P})\cdot\mathbf{W}_v\,\mathbf{A}_w\Big] + (\mathbf{W}_s\cdot\mathbf{P})\,\mathbf{A}_s. \tag{10.9}$$

A way to decompose a vector in terms of the derivatives of $\mathbf{W}$ is

$$\mathbf{F} = \frac{1}{D}\left[(\mathbf{W}_w \times \mathbf{W}_s)(\mathbf{F}\cdot\mathbf{W}_v) - (\mathbf{W}_v \times \mathbf{W}_s)(\mathbf{F}\cdot\mathbf{W}_w)\right] + \mathbf{W}_s(\mathbf{F}\cdot\mathbf{W}_s). \tag{10.10}$$

Finally, we can show that,

$$\nabla\cdot\mathbf{W}_s = D_s/D \tag{10.11}$$

and from Eq. A.18

$$\nabla\times\mathbf{W}_s = 0. \tag{10.12}$$

10.4 Application to the Maxwell Equations

We first calculate the divergence of $\mathbf{E}$ using Eq. 10.7, which, from Eq. 10.1, must equal *zero*

$$\nabla\cdot\mathbf{E} = \frac{1}{D}\left[(\mathbf{W}_w \times \mathbf{W}_s)\cdot\mathbf{E}_v - (\mathbf{W}_v \times \mathbf{W}_s)\cdot\mathbf{E}_w\right] + \mathbf{W}_s\cdot\mathbf{E}_s = 0. \tag{10.13}$$

Note that $\mathbf{W}_s$ is in the direction of propagation and therefore must be perpendicular to both $\mathbf{E}$ and $\mathbf{H}$ so that $\mathbf{E}\cdot\mathbf{W}_s = \mathbf{H}\cdot\mathbf{W}_s = 0$. Since $\mathbf{W}$ is linear in s it follows that $\mathbf{W}_{ss} = 0$ and that,

$$\mathbf{W}_s\cdot\mathbf{E}_s = \mathbf{W}_s\cdot\mathbf{H}_s = 0. \tag{10.14}$$

Then Eq. 10.13, after rearranging terms, becomes,

$$\Big[\mathbf{E}_v \times \mathbf{W}_w - \mathbf{E}_w \times \mathbf{W}_v\Big]\cdot\mathbf{W}_s = 0. \tag{10.15}$$

From Eq. 10.15, there must exist a vector function $\mathbf{V}$ such that,

$$\mathbf{E}_v \times \mathbf{W}_w - \mathbf{E}_w \times \mathbf{W}_v = D(\mathbf{W}_s \times \mathbf{V}_s). \tag{10.16}$$

Multiplying this by $\mathbf{W}_v$ and $\mathbf{W}_w$ yields the two equations,

$$\begin{cases} \mathbf{E}_v\cdot\mathbf{W}_s = -(\mathbf{W}_s \times \mathbf{V}_s)\cdot\mathbf{W}_v \\ \mathbf{E}_w\cdot\mathbf{W}_s = -(\mathbf{W}_s \times \mathbf{V}_s)\cdot\mathbf{W}_w. \end{cases} \tag{10.17}$$

By using $\mathbf{W}_s\cdot\mathbf{E} = 0$ we can see that $\mathbf{E}_v\cdot\mathbf{W}_s = -\mathbf{E}\cdot\mathbf{W}_{vs}$ and $\mathbf{E}_w\cdot\mathbf{W}_s = -\mathbf{E}\cdot\mathbf{W}_{ws}$ so that Eq. 10.17 becomes,

$$\begin{cases} \mathbf{E}\cdot\mathbf{W}_{vs} = (\mathbf{W}_s \times \mathbf{V}_s)\cdot\mathbf{W}_v \\ \mathbf{E}\cdot\mathbf{W}_{ws} = (\mathbf{W}_s \times \mathbf{V}_s)\cdot\mathbf{W}_w. \end{cases} \tag{10.18}$$

Multiplying the first of these by $\mathbf{W}_{ws}$ and the second by $\mathbf{W}_{vs}$, then subtracting we get,

$$\mathbf{E} \times (\mathbf{W}_{vs} \times \mathbf{W}_{ws}) = (\mathbf{W}_s \times \mathbf{V}_s) \cdot \mathbf{W}_w \mathbf{W}_{vs} - (\mathbf{W}_s \times \mathbf{V}_s) \cdot \mathbf{W}_v \mathbf{W}_{ws}. \quad (10.19)$$

Since $\mathbf{W}_s = \mathbf{S}/n = (u,\ v,\ w)/n$, its derivatives are,

$$\begin{cases} \mathbf{W}_{vs} = (-v,\ u,\ 0)/nu \\ \mathbf{W}_{ws} = (-w,\ 0,\ u)/nu, \end{cases} \quad (10.20)$$

then,

$$\mathbf{W}_{vs} \times \mathbf{W}_{ws} = \mathbf{S}/n^2 u = \mathbf{W}_s/nu. \quad (10.21)$$

With Eq. 10.18 and Eq. 10.21 we get,

$$\mathbf{E} \times \mathbf{W}_s = \mathbf{Q}, \quad (10.22)$$

where,

$$\mathbf{Q} = nu\left[(\mathbf{W}_s \times \mathbf{V}_s) \cdot \mathbf{W}_w \mathbf{W}_{vs} - (\mathbf{W}_s \times \mathbf{V}_s) \cdot \mathbf{W}_v \mathbf{W}_{ws}\right]. \quad (10.23)$$

Recall that $\mathbf{W}_s$ is a unit vector so that $\mathbf{W}_s^2 = 1$. The derivative of $\mathbf{W}_s^2$ must vanish, hence,

$$\mathbf{W}_s \cdot \mathbf{W}_{vs} = \mathbf{W}_s \cdot \mathbf{W}_{ws} = 0, \quad (10.24)$$

From this it is evident that $\mathbf{Q} \cdot \mathbf{W}_s = 0$. From Eq. A.21 it can be seen that $\mathbf{Q}$ can also be written as a directional derivative

$$\mathbf{Q} = nuD(\mathbf{V}_s \cdot \nabla)\mathbf{W}_s. \quad (10.25)$$

The vector product of $\mathbf{Q}$ from Eq. 10.23 with $\mathbf{W}_s$ (recall that $\mathbf{E} \cdot \mathbf{W}_s = 0$) yields the expression for $\mathbf{E}$

$$\mathbf{E} = \mathbf{W}_s \times \mathbf{Q}. \quad (10.26)$$

Next we calculate the curl of $\mathbf{E}$, the first equation in Eq. 10.3. From Eq. A.23 we get

$$\nabla \times \mathbf{E} = \frac{1}{D}[(\mathbf{W}_w \times \mathbf{W}_s) \times \mathbf{E}_v - (\mathbf{W}_v \times \mathbf{W}_s) \times \mathbf{E}_w] + \mathbf{W}_s \times \mathbf{E}_s = -\sqrt{\frac{\mu}{\epsilon}}\mathbf{H}_s. \quad (10.27)$$

When expanded this becomes,

$$\begin{aligned} &-\tfrac{1}{D}[(\mathbf{E}_v \cdot \mathbf{W}_s)\mathbf{W}_w - (\mathbf{E}_v \cdot \mathbf{W}_w)\mathbf{W}_s \\ &-(\mathbf{E}_w \cdot \mathbf{W}_s)\mathbf{W}_v + (\mathbf{E}_w \cdot \mathbf{W}_v)\mathbf{W}_s] \\ &+\mathbf{W}_s \times \mathbf{E}_s = -\sqrt{\tfrac{\mu}{\epsilon}}\mathbf{H}_s. \end{aligned} \quad (10.28)$$

Since $\mathbf{H} \cdot \mathbf{W}_s = 0$ the scalar product of Eq. 10.28 with $\mathbf{W}_s$ must vanish. As a consequence,

$$\mathbf{E}_v \cdot \mathbf{W}_w - \mathbf{E}_w \cdot \mathbf{W}_v = 0, \tag{10.29}$$

so that Eq. 10.28 becomes,

$$-\frac{1}{D}[(\mathbf{E}_v \cdot \mathbf{W}_s)\mathbf{W}_w - (\mathbf{E}_w \cdot \mathbf{W}_s)\mathbf{W}_v] + \mathbf{W}_s \times \mathbf{E}_s = -\sqrt{\frac{\mu}{\epsilon}}\mathbf{H}_s. \tag{10.30}$$

First we look at Eq. 10.27 to which we apply Eq. 10.24 to get,

$$(\mathbf{W}_s \times \mathbf{Q})_v \cdot \mathbf{W}_w - (\mathbf{W}_s \times \mathbf{Q})_w \cdot \mathbf{W}_v = 0,$$

which expands to,

$$(\mathbf{W}_{vs} \times \mathbf{Q} + \mathbf{W}_s \times \mathbf{Q}_v) \cdot \mathbf{W}_w - (\mathbf{W}_{ws} \times \mathbf{Q} + \mathbf{W}_s \times \mathbf{Q}_w) \cdot \mathbf{W}_v = 0,$$

which reorganizes itself into,

$$\begin{aligned}(\mathbf{W}_{vs} \times \mathbf{W}_w) \cdot \mathbf{Q} - (\mathbf{W}_v \times \mathbf{W}_{ws}) \cdot \mathbf{Q} \\ - (\mathbf{W}_w \times \mathbf{W}_s) \cdot \mathbf{Q}_v + (\mathbf{W}_v \times \mathbf{W}_s) \cdot \mathbf{Q}_w = 0.\end{aligned}$$

The first two terms are,

$$-(\mathbf{W}_v \times \mathbf{W}_w)_s \cdot \mathbf{Q} = -(D\mathbf{W}_s)_s \cdot \mathbf{Q} = -D_s \mathbf{W}_s \cdot \mathbf{Q} = 0.$$

The second two terms are exactly $\nabla \cdot \mathbf{Q}$ so that it reduces to,

$$\nabla \cdot \mathbf{Q} = 0. \tag{10.31}$$

We consider Eq. 10.28 next and make substitutions from Eqs. 10.16 and 10.25 to get,

$$\begin{aligned}&\frac{1}{D}\Big[(\mathbf{W}_s \times \mathbf{V}_s) \cdot \mathbf{W}_v \mathbf{W}_w - (\mathbf{W}_s \times \mathbf{V}_s) \cdot \mathbf{W}_w \mathbf{W}_v\Big] \\ &\quad + \mathbf{W}_s \times (\mathbf{W}_s \times \mathbf{Q}_s) = -\sqrt{\frac{\mu}{\epsilon}}\mathbf{H}_s,\end{aligned}$$

which reduces to,

$$-\frac{1}{D}\Big[(\mathbf{W}_s \times \mathbf{V}_s) \times (\mathbf{W}_v \times \mathbf{W}_w)\Big] - \mathbf{Q}_s = -\sqrt{\frac{\mu}{\epsilon}}\mathbf{H}_s,$$

that leads to, using Eq. 4.7,

$$\mathbf{H}_s = \sqrt{\frac{\epsilon}{\mu}}\Big[\mathbf{Q}_s + (\mathbf{W}_s \times \mathbf{V}_s) \times \mathbf{W}_s\Big], \tag{10.32}$$

which, on integration, becomes,

$$\mathbf{H} = \sqrt{\frac{\epsilon}{\mu}}\Big[\mathbf{Q} + (\mathbf{W}_s \times \mathbf{V}) \times \mathbf{W}_s\Big] = \sqrt{\frac{\epsilon}{\mu}}\Big[\mathbf{Q} + \mathbf{V} - (\mathbf{W}_s \cdot \mathbf{V})\,\mathbf{W}_s\Big]. \tag{10.33}$$

A vector constant of integration needs to be introduced at this point. However, since its scalar product with $\mathbf{W}_s$ must vanish it can be absorbed into the arbitrary vector function $\mathbf{V}$.

From two of the Maxwell equations we have obtained expressions for the electric field vector, $\mathbf{E}$, and the magnetic field vector, $\mathbf{H}$, in terms of the variables v, w and s, and an arbitrary vector function $\mathbf{V}$. We use the remaining two Maxwell equations for find conditions on $\mathbf{V}$.

But first, an immediate consequence of these results comes from the fact that $\mathbf{E} \cdot \mathbf{H} = 0$ which leads to

$$(\mathbf{W}_s \times \mathbf{V}) \cdot \mathbf{Q} = 0. \tag{10.34}$$

10.5 Conditions on V

Now we return to the fourth equation of Eq. 10.1 and the first equation of 10.3. For both of these we first need to calculate the gradient of $\mathbf{W}_s \cdot \mathbf{V}$ for which we use Eq. A.12 to get

$$\begin{aligned}\nabla(\mathbf{W}_s \cdot \mathbf{V}) &= (\mathbf{W}_s \cdot \nabla)\mathbf{V} + (\mathbf{V} \cdot \nabla)\mathbf{W}_s \\ &\quad + \mathbf{W}_s \times (\nabla \times \mathbf{V}) + \mathbf{V} \times (\nabla \times \mathbf{W}_s) \\ &= \mathbf{V}_s + (\mathbf{V} \cdot \nabla)\mathbf{W}_s + \mathbf{W}_s \times (\nabla \times \mathbf{V}),\end{aligned} \tag{10.35}$$

since the curl of $\mathbf{W}_s$ vanishes. To complete the calculation we need to use Eq. A.21 to get

$$(\mathbf{V} \cdot \nabla)\mathbf{W}_s = \frac{1}{D}\Big[(\mathbf{W}_s \times \mathbf{V}) \cdot \mathbf{W}_w \mathbf{W}_{vs} - (\mathbf{W}_s \times \mathbf{V}) \cdot \mathbf{W}_v \mathbf{W}_{ws}\Big]. \tag{10.36}$$

and Eq. A.23

$$\begin{aligned}\mathbf{W}_s \times (\nabla \times \mathbf{V}) &= \frac{1}{D}\mathbf{W}_s \times \Big[(\mathbf{W}_w \times \mathbf{W}_s) \times \mathbf{V}_v \\ &\quad -(\mathbf{W}_v \times \mathbf{W}_s) \times \mathbf{V}_w + D(\mathbf{W}_s \times \mathbf{V}_s)\Big] \\ &= \frac{1}{D}\mathbf{W}_s \times \Big[- (\mathbf{V}_v \cdot \mathbf{W}_s)\mathbf{W}_w - (\mathbf{V}_v \cdot \mathbf{W}_w)\mathbf{W}_s \\ &\quad +(\mathbf{V}_w \cdot \mathbf{W}_s)\mathbf{W}_v - (\mathbf{V}_w \cdot \mathbf{W}_v)\mathbf{W}_s \\ &\quad +D(\mathbf{W}_s \cdot \mathbf{V}_s\Big] \\ &\frac{1}{D}\Big\{(\mathbf{W}_w \times \mathbf{W}_s)(\mathbf{V}_v \cdot \mathbf{W}_s) - (\mathbf{W}_v \times \mathbf{W}_s)(\mathbf{V}_w \cdot \mathbf{W}_s) \\ &\quad D\Big[(\mathbf{W}_s \cdot \mathbf{V}_s)\mathbf{W}_s - \mathbf{V}_s\Big]\Big\}.\end{aligned} \tag{10.37}$$

The insertion of Eqs. 10.34 and 10.35 into Eq. 10.33 results in

$$\begin{aligned}
\nabla(\mathbf{W}_s \cdot \mathbf{V}) &= \mathbf{V}_s + \frac{1}{D}\Big[(\mathbf{W}_w \times \mathbf{W}_s) \cdot \mathbf{V}\,\mathbf{W}_{vs} \\
&\quad -(\mathbf{W}_v \times \mathbf{W}_s) \cdot \mathbf{V}\,\mathbf{W}_{ws}\Big] \\
&\quad +\frac{1}{D}\Big[(\mathbf{W}_w \times \mathbf{W}_s)(\mathbf{V}_v \cdot \mathbf{W}_s) - (\mathbf{W}_v \times \mathbf{W}_s)(\mathbf{V}_w \cdot \mathbf{W}_s)\Big] \\
&\quad +(\mathbf{W}_s \cdot \mathbf{V}_s)\mathbf{W}_s - \mathbf{V}_s \qquad (10.38) \\
&= \frac{1}{D}\Big[(\mathbf{W}_w \times \mathbf{W}_s) \cdot \mathbf{V}\,\mathbf{W}_{vs} - (\mathbf{W}_v \times \mathbf{W}_s) \cdot \mathbf{V}\,\mathbf{W}_{ws} \\
&\quad +(\mathbf{W}_w \times \mathbf{W}_s)(\mathbf{V}_v \cdot \mathbf{W}_s)\mathbf{V}\mathbf{W}_{vs} \\
&\quad -(\mathbf{W}_v \times \mathbf{W}_s)(\mathbf{V}_w \cdot \mathbf{W}_s)\Big] + (\mathbf{W}_s \cdot \mathbf{V}_s)\mathbf{W}_s.
\end{aligned}$$

Now refer back to Eq. 10.31 from which we calculate the divergence of $\mathbf{H}$

$$\nabla \cdot \mathbf{H} = \sqrt{\frac{\epsilon}{\mu}}\big[\mathbf{Q} + \mathbf{V} - \nabla(\mathbf{W}_s \cdot \mathbf{V}) \cdot \mathbf{W}_s - (\mathbf{W}_s \cdot \mathbf{V})(\nabla \cdot \mathbf{W}_s)\big] = 0 \qquad (10.39)$$

from the fourth equation of 10.1. From Eq. 10.31 the divergence of $\mathbf{Q}$ vanishes so that we may write

$$\begin{aligned}
\nabla \cdot \mathbf{V} = \frac{1}{D}\mathbf{W}_s \cdot \Big[&(\mathbf{W}_w \times \mathbf{W}_s) \cdot \mathbf{V}\,\mathbf{W}_{vs} - (\mathbf{W}_v \times \mathbf{W}_s) \cdot \mathbf{V}\,\mathbf{W}_{ws} \\
&+(\mathbf{W}_w \times \mathbf{W}_s)(\mathbf{V}_v \cdot \mathbf{W}_s) - (\mathbf{W}_v \times \mathbf{W}_s)(\mathbf{V}_w \cdot \mathbf{W}_s) \qquad (10.40) \\
&+D(\mathbf{W}_s \cdot \mathbf{V}_s)\mathbf{W}_s + D_s(\mathbf{W}_s \cdot \mathbf{V})\mathbf{W}_s\Big]
\end{aligned}$$

which reduces further to

$$D(\nabla \cdot \mathbf{V}) = D(\mathbf{W}_s \cdot \mathbf{V}_s) + D_s(\mathbf{W}_s \cdot \mathbf{V}) = \Big[D(\mathbf{W}_s \cdot \mathbf{V})\Big]_s. \qquad (10.41)$$

Next refer to the second equation of Eq. 10.3 which relates the curl of $\mathbf{H}$ to the derivative of $\mathbf{E}$. From Eq. 10.26 we can calculate

$$\nabla \times \mathbf{H} = \sqrt{\frac{\epsilon}{\mu}}\Big[\nabla \times (\mathbf{Q} + \mathbf{V}) - \nabla(\mathbf{W}_s \cdot \mathbf{V}) \times \mathbf{W}_s - (\mathbf{W}_s \cdot \mathbf{V})(\nabla \times \mathbf{W}_s)\Big]. \qquad (10.42)$$

In Appendix Eq. A.34 it is shown that $\nabla \times \mathbf{W}_s = 0$. We use Eq. 10.36 to calculate

$$
\begin{aligned}
&-\nabla(\mathbf{W}_s \cdot \mathbf{V}) \times \mathbf{W}_s = +\frac{1}{D}\mathbf{W}_s \times \Big[(\mathbf{W}_w \times \mathbf{W}_s) \cdot \mathbf{V}\,\mathbf{W}_{vs} \\
&-(\mathbf{W}_v \times \mathbf{W}_s) \cdot \mathbf{V}\,\mathbf{W}_{ws} + (\mathbf{W}_w \times \mathbf{W}_s)(\mathbf{V}_v \cdot \mathbf{W}_s) \\
&-(\mathbf{W}_v \times \mathbf{W}_s)(\mathbf{V}_w \cdot \mathbf{W}_s) + (\mathbf{W}_s \cdot \mathbf{V}_s)\mathbf{W}_s\Big] \qquad (10.43)\\
&= \frac{1}{D}\Big[(\mathbf{W}_w \times \mathbf{W}_s) \cdot \mathbf{V}(\mathbf{W}_s \times \mathbf{W}_{vs}) - (\mathbf{W}_v \times \mathbf{W}_s) \cdot \mathbf{V}(\mathbf{W}_s \times \mathbf{W}_{ws}) \\
&+(\mathbf{V}_v \cdot \mathbf{W}_s)\mathbf{W}_w - (\mathbf{V}_w \cdot \mathbf{W}_s)\mathbf{W}_v,
\end{aligned}
$$

so that

$$
\begin{aligned}
\nabla \times \mathbf{H} \quad &= \sqrt{\frac{\epsilon}{\mu}}\Big\{\nabla \times (\mathbf{Q} + \mathbf{V}) \\
+\frac{1}{D}\Big[(\mathbf{W}_w \times \mathbf{W}_s) &\cdot \mathbf{V}(\mathbf{W}_s \times \mathbf{W}_v)_s - (\mathbf{W}_v \times \mathbf{W}_s) \cdot \mathbf{V}(\mathbf{W}_s \times \mathbf{W}_w)_s \\
&+(\mathbf{V}_v \cdot \mathbf{W}_s)\mathbf{W}_w - (\mathbf{V}_w \cdot \mathbf{W}_s)\mathbf{W}_v\Big]\Big\} \qquad (10.44)\\
= \sqrt{\frac{\epsilon}{\mu}}\mathbf{E}_s \quad &= \sqrt{\frac{\epsilon}{\mu}}(\mathbf{W}_s \times \mathbf{Q}_s).
\end{aligned}
$$

This reduces to

$$
\begin{aligned}
&-(\mathbf{Q}_v \cdot \mathbf{W}_s)\mathbf{W}_w + (\mathbf{Q}_v \cdot \mathbf{W}_w)\mathbf{W}_s \\
&+(\mathbf{Q}_w \cdot \mathbf{W}_s)\mathbf{W}_v - (\mathbf{Q}_w \cdot \mathbf{W}_v)\mathbf{W}_s \\
&+(\mathbf{V}_v \cdot \mathbf{W}_w)\mathbf{W}_s - (\mathbf{V}_w \cdot \mathbf{W}_v)\mathbf{W}_s + D(\mathbf{W}_s \times \mathbf{V}_s) \qquad (10.45)\\
&+(\mathbf{W}_w \times \mathbf{W}_s) \cdot \mathbf{V}(\mathbf{W}_s \times \mathbf{W}_v)_s \\
&-(\mathbf{W}_v \times \mathbf{W}_s) \cdot \mathbf{V}(\mathbf{W}_s \times \mathbf{W}_w)_s = 0.
\end{aligned}
$$

The scalar product of this with $\mathbf{W}_s$ results in

$$(\mathbf{Q}_v \cdot \mathbf{W}_w) - (\mathbf{Q}_w \cdot \mathbf{W}_v) + (\mathbf{V}_v \cdot \mathbf{W}_w) - (\mathbf{V}_w \cdot \mathbf{W}_v) = 0 \qquad (10.46)$$

so that Eq. 10.43 becomes

$$
\begin{aligned}
&-(\mathbf{Q}_v \cdot \mathbf{W}_s)\mathbf{W}_w + (\mathbf{Q}_w \cdot \mathbf{W}_s)\mathbf{W}_v + D(\mathbf{W}_s \times \mathbf{V}_s) \\
&+(\mathbf{W}_w \times \mathbf{W}_s) \cdot \mathbf{V}(\mathbf{W}_s \times \mathbf{W}_v)_s \qquad (10.47)\\
&-(\mathbf{W}_v \times \mathbf{W}_s) \cdot \mathbf{V}(\mathbf{W}_s \times \mathbf{W}_w)_s = 0
\end{aligned}
$$

What we have here are four equations, Eqs. 10.32, 10.39, 10.44 and 10.45 that impose conditions on the vector function $\mathbf{V}$.

From this and the fact that $\mathbf{W}_s \cdot \mathbf{Q} = 0$ indicates that $\mathbf{Q}$ is perpendicular to both $\mathbf{W}_s$ and $\mathbf{W}_s \times \mathbf{V}$ and then must be parallel to their vector product. Therefore there must exist an α such that,

$$\mathbf{Q} = \alpha(\mathbf{W}_s \times \mathbf{V}) \times \mathbf{W}_s = \alpha[\mathbf{V} - (\mathbf{V} \cdot \mathbf{W}_s)\mathbf{W}_s]. \tag{10.48}$$

In these calculations we have used Eq. A.6. In what follows we make use of Eqs. 10.7 and 10.11.

To Eq. 10.48 we apply Eq. 10.41 to get,

$$\begin{aligned}\nabla \cdot \mathbf{Q} &= \nabla\alpha \cdot [\mathbf{V} - (\mathbf{W}_s \cdot \mathbf{V})\mathbf{W}_s] \\ &\quad +\alpha[\nabla \cdot \mathbf{V} - \nabla(\mathbf{W}_s \cdot \mathbf{V}) \cdot \mathbf{W}_s(\mathbf{W}_s \cdot \mathbf{V})D_s/D \\ &= \nabla\alpha \cdot \mathbf{V} - (\mathbf{W}_s \cdot \mathbf{V})\alpha_s + \alpha(\nabla \cdot \mathbf{V}) \\ &\quad -\alpha(\mathbf{W}_s \cdot \nabla)(\mathbf{W}_s \cdot \mathbf{V}) - \alpha(\mathbf{W}_s \cdot \mathbf{V})D_s/D.\end{aligned} \tag{10.49}$$

Rearranging this yields,

$$D(\nabla \cdot \mathbf{Q}) = D\nabla \cdot \alpha\mathbf{V} - [D\,\alpha\,(\mathbf{W}_s \cdot \mathbf{V})]_s = 0, \tag{10.50}$$

so that,

$$\nabla \cdot (\alpha\,\mathbf{V}) = (1/D)\,[D\,\alpha\,(\mathbf{W}_s \cdot \mathbf{V})]_s. \tag{10.51}$$

Next we apply Eq. 10.48 to Eq. 10.26 to get an expression for $\mathbf{E}$ in terms of $\mathbf{V}$, which yields,

$$\mathbf{E} = \alpha\,(\mathbf{W}_s \times \mathbf{V}). \tag{10.52}$$

We do the same thing for $\mathbf{H}$ by applying Eq. 10.48 to Eq. 10.33,

$$\mathbf{H} = \sqrt{\frac{\epsilon}{\mu}}[\mathbf{Q} + (\mathbf{W}_s \times \mathbf{V}) \times \mathbf{W}_s] = \sqrt{\frac{\epsilon}{\mu}}\,(\alpha+1)\,(\mathbf{W}_s \times \mathbf{V}) \times \mathbf{W}_s, \tag{10.53}$$

from which it is clear that $\mathbf{E} \cdot \mathbf{H} = 0$, just as it should be. And the Poynting vector becomes,

$$\mathbf{P} = -\frac{c}{4\pi}\sqrt{\frac{\epsilon}{\mu}}\alpha\,(\alpha+1)\,(\mathbf{V} \times \mathbf{W}_s)^2\mathbf{W}_s. \tag{10.54}$$

The divergence of $\mathbf{H}$, from Eq. 10.1, equals zero. Applying this to Eq. 10.53 yields,

$$\begin{aligned}\nabla \cdot \mathbf{H} &= \sqrt{\frac{\epsilon}{\mu}}\,\{\nabla\,(\alpha+1) \cdot [\mathbf{V} - (\mathbf{W}_s \cdot \mathbf{V})\mathbf{W}_s] \\ &\quad +(\alpha+1)\,[\nabla \cdot \mathbf{V} - \nabla(\mathbf{W}_s \cdot \mathbf{V}) \cdot \mathbf{W}_s - \mathbf{W}_s \cdot \mathbf{V})(\nabla \cdot \mathbf{W}_s)]\} \\ &= \sqrt{\tfrac{\epsilon}{\mu}}\,\{\nabla\,(\alpha{+}1) \cdot \mathbf{V} + (\alpha{+}1)\,\nabla \cdot \mathbf{V} - (\mathbf{W}_s \cdot \mathbf{V})\,(\alpha{+}1)_s \\ &\quad -(\alpha+1)\,(\mathbf{W}_s \cdot \mathbf{V}_s) - (\alpha+1)\,(\mathbf{W}_s \cdot \mathbf{V})\frac{D_s}{D}\Big\} = 0.\end{aligned} \tag{10.55}$$

From this comes,

$$\nabla \cdot [(\alpha+1)\,\mathbf{V}] = \frac{1}{D}\,[D\,(\alpha+1)\,(\mathbf{W}_s \cdot \mathbf{V})]_s \,. \tag{10.56}$$

Subtracting from this Eq. 10.51 yields,

$$D(\nabla \cdot \mathbf{V}) = [D(\mathbf{V} \cdot \mathbf{W}_s)]_s, \tag{10.57}$$

which leads us to,

$$\alpha = 1. \tag{10.58}$$

Now the expressions for **E**, **H** and **P** become, from Eqs. 10.52, 10.53 and 10.54,

$$\mathbf{E} = (\mathbf{W}_s \times \mathbf{V}), \tag{10.59}$$

$$\mathbf{H} = 2\sqrt{\frac{\epsilon}{\mu}}(\mathbf{W}_s \times \mathbf{V}) \times \mathbf{W}_s, \tag{10.60}$$

and,

$$\mathbf{P} = -2\frac{c}{4\pi}\sqrt{\frac{\epsilon}{\mu}}(\mathbf{V} \times \mathbf{W}_s)^2\mathbf{W}_s. \tag{10.61}$$

Finally we come to the second equation in Eq. 10.3, $\nabla \times \mathbf{H} = \sqrt{\epsilon/\mu}\,\mathbf{E}_s$, which becomes, after applying Eq. 10.5 to **H** in Eq. 10.44 and using the derivative of **E** from Eq. 10.45,

$$2\{\nabla \times \mathbf{V} - \nabla(\mathbf{V} \cdot \mathbf{W}_s) \times \mathbf{W}_s\} = \mathbf{W}_s \times \mathbf{V}_s,$$

which becomes after some calculations and reductions,

$$2\{\nabla \times \mathbf{V} + \mathbf{W}_s \times [(\mathbf{V} \cdot \nabla)\mathbf{W}_s + \mathbf{V}_s + \mathbf{W}_s \times (\nabla \times \mathbf{V})]\} = \mathbf{W}_s \times \mathbf{V}_s.$$

Continuing further we finally arrive at,

$$\mathbf{W}_s \times \mathbf{V}_s + 2\{\mathbf{W}_s \times [(\mathbf{V} \cdot \nabla)\mathbf{W}_s] + \mathbf{W}_s \cdot (\nabla \times \mathbf{V})\,\mathbf{W}_s\} = 0. \tag{10.62}$$

Taking the scalar product of this with $\mathbf{W}_s$ we get,

$$\mathbf{W}_s \cdot (\nabla \times \mathbf{V}) = 0. \tag{10.63}$$

When this is substituted into Eq. 10.62 we get,

$$\mathbf{W}_s \times \mathbf{V}_s + 2\,\mathbf{W}_s \times [(\mathbf{V} \cdot \nabla)\mathbf{W}_s] = 0. \tag{10.64}$$

10.6 Conditions on the Vector V

We have found three conditions that must be applied to the vector function $\mathbf{V}$ and its derivatives. These are given in Eqs. 10.57, 10.63 and 10.64. We will look at Eq. 10.24 first to which we apply the expression for divergence given in Eq. A.17. The result is,

$$(\mathbf{W}_w \times \mathbf{W}_s) \cdot \mathbf{V}_v - (\mathbf{W}_v \times \mathbf{W}_s) \cdot \mathbf{V}_w = D_s(\mathbf{V} \cdot \mathbf{W}_s). \tag{10.65}$$

To Eq. 10.63 we apply the expression for the curl in Eq. A.23 which results in,

$$\mathbf{W}_w \cdot \mathbf{V}_v - \mathbf{W}_v \cdot \mathbf{V}_w = 0. \tag{10.66}$$

Finally we use Eq. A.26, the expression for the directional derivative, to reduce Eq. 10.64. From that calculation we get,

$$\begin{aligned} D(\mathbf{W}_s \times \mathbf{V}_s) + 2[(\mathbf{W}_s \times \mathbf{V}) \cdot \mathbf{W}_w(\mathbf{W}_s \times \mathbf{W}_v)_s \\ -(\mathbf{W}_s \times \mathbf{V}) \cdot \mathbf{W}_v(\mathbf{W}_s \times \mathbf{W}_w)_s] = 0. \end{aligned} \tag{10.67}$$

So, for the expressions for $\mathbf{E}$ and $\mathbf{H}$, given in Eqs. 10.59 and 10.60, respectively to satisfy all four Maxwell equations in Eqs. 10.1 and 10.3 as well as the side condition $\mathbf{E} \cdot \mathbf{H} = 0$ the vector function $\mathbf{V}$ must satisfy Eqs. 10.65, 10.66 and 10.67.

10.7 Spherical Wavefronts

The expressions that we have obtained for $\mathbf{E}$ and $\mathbf{H}$ contain two arbitrary functions. In addition to the k-function, in which resides the geometric aberrations of the wavefront train there is the vector function $\mathbf{V}$ which appears to be determined by the physical properties of the electromagnetic radiation. In this section we restrict ourselves to spherical wavefronts, thus determining completely the k-function, and applying the results to the three vector equations that involve $\mathbf{V}$.

In Chapter 6 we obtained the form of the vector function $\mathbf{W}(v,\ w,\ s)$ when the wavefront train consists of concentric spherical waves. With the center of these wavefronts located at the point $-(0,\ h,\ t)$ we got,

$$\mathbf{W}_w(v,\ w,\ s) = \frac{s}{n}\mathbf{S} - \mathbf{K}, \tag{6.27}$$

where,

$$k = v\,h + w\,t, \tag{6.26}$$

and,

$$k_v = h, \qquad k_w = t, \tag{6.25}$$

so that,

$$\mathbf{K} = (0,\ h,\ t). \tag{10.68}$$

We also found the partial derivatives,

$$\mathbf{W}_v = -\frac{s}{n\,u}(\mathbf{S}\times\mathbf{Z}), \qquad \mathbf{W}_w = \frac{s}{n\,u}(\mathbf{S}\times\mathbf{Y}), \tag{6.28}$$

the differential element of area,

$$D = \frac{s^2}{n\,u}, \tag{6.30}$$

and its derivative,

$$D_s = \frac{2\,s}{n\,u}.$$

In getting Eq. 6.28 we used,

$$\begin{cases} \mathbf{S}_v = -\dfrac{1}{u}(v, -u,\ 0) = -\dfrac{1}{u}(\mathbf{S}\times\mathbf{Z}) \\ \mathbf{S}_w = -\dfrac{1}{u}(w,\ 0, -u) = \ \dfrac{1}{u}(\mathbf{S}\times\mathbf{Y}), \end{cases} \tag{6.2}$$

which we will need in this section.

These results we will apply first to Eq. 10.57 which we rearrange in the form,

$$\begin{aligned} D(\mathbf{W}_s\times\mathbf{V})_s = {} & (\mathbf{W}_w\times\mathbf{W}_s)\cdot\mathbf{V}(\mathbf{W}_v\times\mathbf{W}_s)_s \\ & -(\mathbf{W}_v\times\mathbf{W}_s)\cdot\mathbf{V}(\mathbf{W}_w\times\mathbf{W}_s)_s. \end{aligned} \tag{10.69}$$

We use Eq. 6.28 to make the following calculations,

$$\begin{cases} \mathbf{W}_w\times\mathbf{W}_s = \dfrac{s}{n^2u}(\mathbf{S}\times\mathbf{Y})\times\mathbf{S} = \dfrac{-s}{n^2u}(v\mathbf{S}-n^2\mathbf{Y}) \\ \mathbf{W}_v\times\mathbf{W}_s = \dfrac{-s}{n^2u}(\mathbf{S}\times\mathbf{Z})\times\mathbf{S} = \dfrac{s}{n^2u}(w\mathbf{S}-n^2\mathbf{Z}), \end{cases} \tag{10.70}$$

which we substitute into Eq. 10.57 to get,

$$\begin{aligned} D(\mathbf{W}_s\times\mathbf{V})_s &= \frac{-s}{n^4u^2}[(v\mathbf{S}-n^2\mathbf{Y})\cdot\mathbf{V}(w\mathbf{S}-n^2\mathbf{Z}) \\ &\qquad -(w\mathbf{S}-n^2\mathbf{Z})\cdot\mathbf{V}(v\mathbf{S}-n^2\mathbf{Y})] \\ &= \frac{-s}{n^2u^2}\{(\mathbf{S}\cdot\mathbf{V})(w\mathbf{Y}-v\mathbf{Z})-[\mathbf{V}\cdot(w\mathbf{Y}-v\mathbf{Z})]\mathbf{S} \\ &\qquad +n^2[(\mathbf{V}\cdot\mathbf{Y})\mathbf{Z}-(\mathbf{V}\cdot\mathbf{Z})\mathbf{Y}]\} \\ &= \frac{s}{n^2u^2}\{\mathbf{V}\times[\mathbf{S}\times(w\mathbf{Y}-v\mathbf{Z})]+n^2\mathbf{V}\times(\mathbf{Y}\times\mathbf{Z})\} \\ &= \frac{s}{n^2u^2}\mathbf{V}\times[w(\mathbf{S}\times\mathbf{Y})-v(\mathbf{S}\times\mathbf{Z})+n^2\mathbf{X}]. \end{aligned} \tag{10.71}$$

Now we use Eq. 6.2 to break this up into its vector components,

$$\begin{aligned} D(\mathbf{W}_s \times \mathbf{V})_s &= \frac{s}{n^2u^2}\mathbf{V} \times [w(-w,\ 0,\ u) \\ &\quad -v(v,\ -u\ ,0) + n^2(1,\ 0,\ 0)] \\ &= \frac{s}{n^2u^2}\mathbf{V} \times (n^2 - v^2 - w^2,\ uv, uw) \\ &= \frac{s}{n^2u}\mathbf{V} \times \mathbf{S}. \end{aligned} \tag{10.72}$$

Now we use Eq. 6.30 to get,

$$s^2(\mathbf{W}_s \times \mathbf{V})_s + 2s(\mathbf{W}_s \times \mathbf{V}) = 0, \tag{10.73}$$

an ordinary differential equation whose solution is,

$$s^2(\mathbf{W}_s \times \mathbf{V}) = \mathbf{A}, \qquad \mathbf{A}_s = 0. \tag{10.74}$$

From Eq. 10.59 we can see that,

$$\mathbf{E} = \frac{1}{s^2}\mathbf{A}, \tag{10.75}$$

the law of least squares.

Next we come to Eq. 10.65 which we rearrange as,

$$D_s(\mathbf{V} \cdot \mathbf{W}_s) = \mathbf{W}_w \cdot (\mathbf{W}_s \times \mathbf{V}_v) - \mathbf{W}_v \cdot (\mathbf{W}_s \times \mathbf{V}_w). \tag{10.76}$$

First note that,

$$(\mathbf{W}_s \times \mathbf{V})_v = \mathbf{W}_s \times \mathbf{V}_v + \mathbf{W}_{vs} \times \mathbf{V} = \frac{1}{s^2}\mathbf{A}_v,$$

with a similar expression for the derivative with respect to w. These lead to,

$$\begin{cases} \mathbf{W}_s \times \mathbf{V}_v = \dfrac{1}{s^2}\mathbf{A}_v - \mathbf{W}_{vs} \times \mathbf{V} \\ \mathbf{W}_s \times \mathbf{V}_w = \dfrac{1}{s^2}\mathbf{A}_w - \mathbf{W}_{ws} \times \mathbf{V} \end{cases} \tag{10.77}$$

Doing the appropriate multiplications we find,

$$\begin{cases} \mathbf{W}_w \cdot (\mathbf{W}_s \times \mathbf{V}_v) = \dfrac{1}{s^2}\mathbf{A}_v \cdot \mathbf{W}_w - (\mathbf{W}_w \times \mathbf{W}_{vs}) \cdot \mathbf{V} \\ \mathbf{W}_v \cdot (\mathbf{W}_s \times \mathbf{V}_w) = \dfrac{1}{s^2}\mathbf{A}_w \cdot \mathbf{W}_v - (\mathbf{W}_v \times \mathbf{W}_{ws}) \cdot \mathbf{V}. \end{cases} \tag{10.78}$$

Substituting this back into Eq. 10.76 yields the following,

$$\begin{aligned} D_s(\mathbf{V}\cdot\mathbf{W}_s) - \frac{1}{s^2}[(\mathbf{A}_v\cdot\mathbf{W}_w) - (\mathbf{A}_w\cdot\mathbf{W}_v)] &= -(\mathbf{W}_w\times\mathbf{W}_{vs})\cdot\mathbf{V} + (\mathbf{W}_v\times\mathbf{W}_{ws})\cdot\mathbf{V} \\ &= (\mathbf{W}_v\times\mathbf{W}_w)_s\cdot\mathbf{V} \\ &= D_s(\mathbf{W}_s\cdot\mathbf{V}). \end{aligned}$$

The final result is,

$$(\mathbf{A}_v\cdot\mathbf{W}_w) - (\mathbf{A}_w\cdot\mathbf{W}_v) = 0, \tag{10.79}$$

which, on the application of Eq. 6.28, becomes,

$$(\mathbf{S}\times\mathbf{Z})\cdot\mathbf{A}_v + (\mathbf{S}\times\mathbf{Y})\cdot\mathbf{A}_w = 0. \tag{10.80}$$

Finally we turn to Eq. 10.66 which, when $\mathbf{W}_v$ and $\mathbf{W}_w$ are substituted from Eq. 6.28 results in,

$$(\mathbf{S}\times\mathbf{Y})\cdot\mathbf{V}_v + (\mathbf{S}\times\mathbf{Z})\cdot\mathbf{V}_w = 0,$$

which we rearrange as,

$$(\mathbf{V}_v\times\mathbf{W}_s)\cdot\mathbf{Y} + (\mathbf{V}_w\times\mathbf{W}_s)\cdot\mathbf{Z} = 0$$

or,

$$(\mathbf{W}_s\times\mathbf{V}_v)\cdot\mathbf{Y} + (\mathbf{W}_s\times\mathbf{V}_w)\cdot\mathbf{Z} = 0. \tag{10.81}$$

Into this we substitute from Eq. 10.77 to get,

$$\begin{aligned} \frac{1}{s^2}[(\mathbf{A}_v\cdot\mathbf{Y}) + (\mathbf{A}_w\cdot\mathbf{Z})] &= \mathbf{V}\cdot[(\mathbf{Y}\times\mathbf{W}_{vs}) + (\mathbf{Z}\times\mathbf{W}_{ws})] \\ &= \frac{1}{n}\mathbf{V}\cdot[(\mathbf{Y}\times\mathbf{S}_v) + (\mathbf{Z}\times\mathbf{S}_w)] \\ &= -\frac{1}{nu}\mathbf{V}\cdot[\mathbf{Y}\times(\mathbf{S}\times\mathbf{Z}) + \mathbf{Z}\times(\mathbf{S}\times\mathbf{Y})] \\ &= \frac{1}{nu}\mathbf{V}\cdot[(\mathbf{S}\cdot\mathbf{Z})\mathbf{Y} - (\mathbf{S}\cdot\mathbf{Y})\mathbf{Z}] \\ &= \frac{1}{nu}\mathbf{V}\cdot[\mathbf{S}\times(\mathbf{Y}\times\mathbf{Z})], \end{aligned}$$

which is the same as,

$$\frac{u}{s^2}[(\mathbf{A}_v\cdot\mathbf{Y}) + (\mathbf{A}_w\cdot\mathbf{Z})] = (\mathbf{V}\times\mathbf{W}_{s,})\cdot\mathbf{X}. \tag{10.82}$$

The ultimate result comes from substituting from Eq. 10.74 to get

$$u[(\mathbf{A}_v \cdot \mathbf{Y}) + (\mathbf{A}_w \cdot \mathbf{Z})] + \mathbf{A} \cdot \mathbf{X} = 0. \tag{10.83}$$

We have found expressions for **E** and **H** as well as the Poynting vector for homogeneous, isotropic media. We did this by using a system of generalized coordinates in terms of which the Maxwell equations are cast. The generalized coordinates are the parameters of wavefront trains, expressions that are a general solution of the eikonal equation. Associated with this is an arbitrary function, the k-function, that arises from the method used in getting the eikonal equation's general solution.

In obtaining solutions for **E** and **H** from the Maxwell equations a second arbitrary function, the vector function **V**, is introduced. **V** and its derivatives are subject to three conditions that assure that **E** and **H** do indeed satisfy the Maxwell equations.

In the k-function resides only information regarding the geometry of the wavefront train. The vector function **V** contains information regarding its physical properties; amplitude, state of polarization, phase relations, etc. One would expect **V** to involve the complex exponential function dependent on the parameter s as well as the k-function.

These solutions depend on two arbitrary functions. Recall that if either **E** or **H** is eliminated from the four Maxwell equations a second order partial differential equation, the wave equation, results. A general solution of such an equation must involve two arbitrary functions.

The whole point of this exercise was to obtain expressions for **E** and **H** within the context of the k-function. These vectors can be associated with points on a wavefront or on the exit pupil of a lens over which a diffraction integral can be calculated. Since there is no restriction on the magnitude of angles these calculations could lead to a vector diffraction theory.

Part III

Ramifications

The Mathematics of Geometrical and Physical Optics: The k-function and its Ramifications. O.N. Stavroudis

ISBN: 3-527-40448-1

11 The Modern Schiefspiegler

The modern schiefspiegler is a wildly off-axis, reflecting optical system usually conceived as a telescope, but with other configurations possible. Its advantage is that it can be designed without a central obscuration and it can be folded to fit into a limited volume such as that found in the restricted space of a satellite. Its disadvantage is that its reflecting surfaces need to be off-axis segments of conic sections of revolution, making fabrication, testing and alignment extremely difficult.

11.1 Background

The history of off-axis telescopes is long indeed. Kutter pointed out that attempts at such designs began as early as 1685 though none was taken seriously until the 1870's. From that point on, until the 1930's, a number of these designs were fabricated. Some were clearly embarrassments; others, mainly those fabricated by Kutter himself, were moderately successful. A summary of Kutter's work appears in Stavroudis.[1]

Also in Kutter[2] appears an outline of subsequent work on these off-axis systems. Beginning with the study of the effects of small tilts on the components on the Seidel aberrations, or, more accurately, their effects on the coefficients of a power series that is an analog of the Seidel terms. Their purpose was to predict the effect of such tilts and displacements on these aberrations in order to establish tolerances on the fabrication of precision optical systems. The extension of these results to the design of off-axis systems was a logical consequence. A detailed examination of these general methods and their subsequent application is presented in a dissertation by Richard Buchroeder[3] that contains a comprehensive bibliography.

Over sixty years ago Anton Kutter gave an account of his experiments in the design of off-axis telescopes.[4] His approach was to modify existing designs, such as the Cassegrain and the Gregorian, by restricting the working area of the primary mirror and then by rotating and translating the secondary mirror with respect to the primary. In this way he could correct residual aberrations. Changes in the power of the secondary were also made. This system he called the *Schiefspiegler*.

The modern schiefspiegler differs from Kutter's original concept in that it consists of two prolate spheroid mirrors coupled by conjoining the distal focus of the first with the proximal focus of the second. A ray through the proximal focus of the first spheroid must, after reflection,

[1] Stavroudis and Ames 1992.
[2] Kutter 1953.
[3] Buchroeder 1970.
[4] Kutter 1969.

The Mathematics of Geometrical and Physical Optics: The k-function and its Ramifications. O.N. Stavroudis

ISBN: 3-527-40448-1

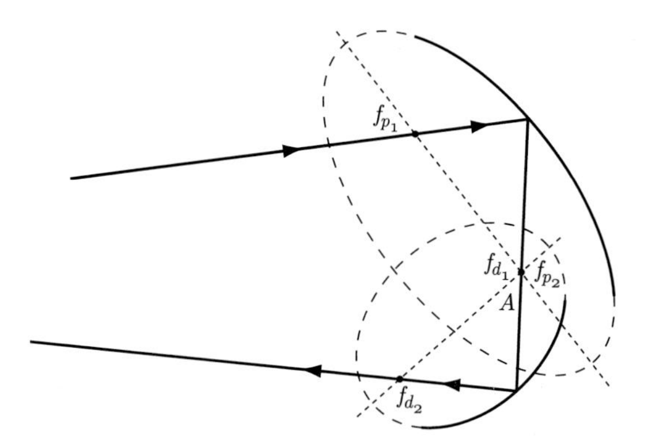

Figure 11.1: *Two Coupled Spheroids.* f_{p_1} is proximal focus of the first ellipsoid, f_{d_2} is the distal focus of the second ellipsoid. A is the location of the distal focus of the first ellipsoid, f_{d_1}, and the proximal focus of the second, f_{p_2}.

pass through the common focus and then, after a second reflection, through the distal focus of the second spheroid. The *pseudo axis* is one such ray with the property that it is a line of symmetry for all other meridional rays. The entrance pupil is located at the first proximal focus; the entrance pupil plane is perpendicular to the pseudo axis. It follows that the exit pupil is located at the second distal focus. Therefore all rays that pass through the foci, in the language of conventional optical systems, are chief rays.

Now arrange two such spheroids so that the distal focus of the first coincides with the proximal focus of the second. Let β be the angle between the two spheroid axes; these, in turn, determine a plane which we will call the *plane of symmetry*. It follows that any ray through the proximal focus of the first spheroid must also, after reflection, pass through the common focus and then, after a second reflection, through the distal focus of the second spheroid.

We locate the system's entrance pupil at the proximal focus of the first spheroid; the exit pupil must then be at the distal focus of the second as is shown in Fig. 11.1. Rays through these foci, in the language of conventional optical systems, are therefore *chief rays*.

At this point we define the *pseudo axis* as that chief ray lying in the plane of symmetry about which all other chief rays that lie in that plane are symmetric. Of course all chief rays not lying on the plane of symmetry will be symmetric with respect to that plane. The condition for the existence of the pseudo axis appears in reference [1]. A shorter proof will be given here in a subsequent section. As part of the definition of the modern schiefspiegler we will take the entrance pupil plane to be perpendicular to the pseudo axis at the first proximal focus; the exit pupil, also normal to the pseudo axis at the second distal focus.

With the pseudo axis established the system possesses many of the properties of a conventional, rotationally symmetric optical system. It is possible to speak of angular magnification[5] and distortion. An unexpected result, quite contrary to intuition, is that magnification is uniform in all directions.

[5] Stavroudis and Flores-Hernandez 1998.

This schiefspiegler differs from the earlier version in that it begins with an off-axis system, imposes conditions and restrictions, and analyzes the resulting structure. There are no approximations; all calculations are exact.

In recognition of Anton Kutter's earlier work I have called this system the *modern schiefspiegler*. Its elements are prolate spheroids, ellipsoids of revolution with the major axis of the generating ellipse as the axis of rotation on which lie the two foci. Earlier work along these lines can be found in reference [1].

11.2 The Single Prolate Spheroid

This is one of the conic sections of revolution that we had described in Chapters 2 and 4 where we choose a focus as a coordinate origin and the axis of rotation as the polar axis. This is shown in Fig. 11.2. As in previous chapters we refer to the focus at the coordinate origin as the spheroid's *proximal focus*; the other as its *distal focus*. In Chapter 4 we applied generalized ray tracing to get expressions for reflected rays and principal directions and principal curvatures for wavefronts associated with rays passing through the proximal and distal foci. As before we use the direction cosine vector, $\mathbf{S} = (\xi,\ \eta,\ \zeta)$. Referring back to Chapter 7, for such a ray we found that the point of incidence is

$$\mathbf{P} = \varrho(\xi,\ \eta,\ \zeta), \tag{7.63}$$

where we obtained,

$$\varrho = \frac{r}{1 - \epsilon\zeta}, \tag{7.64}$$

the distance along the ray from the proximal focus to the point of incidence. Also from Chapter 7 we have the unit normal vector to the reflecting surface at the point of incidence,

$$\mathbf{N} = \frac{1}{\mathcal{K}}(\xi,\ \eta,\ \zeta - \epsilon), \tag{7.65}$$

where,

$$\mathcal{K}^2 = 1 + \epsilon^2 - 2\epsilon\,\zeta. \tag{7.66}$$

The cosine of the angle of incidence was found to be,

$$\cos i = \frac{1 - \epsilon\zeta}{\mathcal{K}}. \tag{7.68}$$

From the ray tracing equation for reflection we have,

$$\mathbf{S}' = \frac{1}{\mathcal{K}}\left((1 - \epsilon^2)\xi,\ (1 - \epsilon^2)\eta,\ (1 + \epsilon^2)\zeta - 2\,\epsilon\right). \tag{7.70}$$

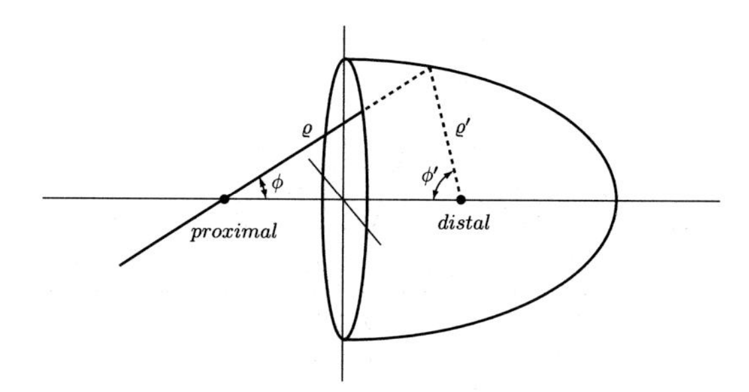

Figure 11.2: *The Prolate Spheroid.* A ray from the proximal focus is reflected by the spheroid and then passes through its distal focus. ϕ and ϕ' are the angles the ray makes with the spheroid axis. ρ and ρ' are the ray path lengths before and after reflection.

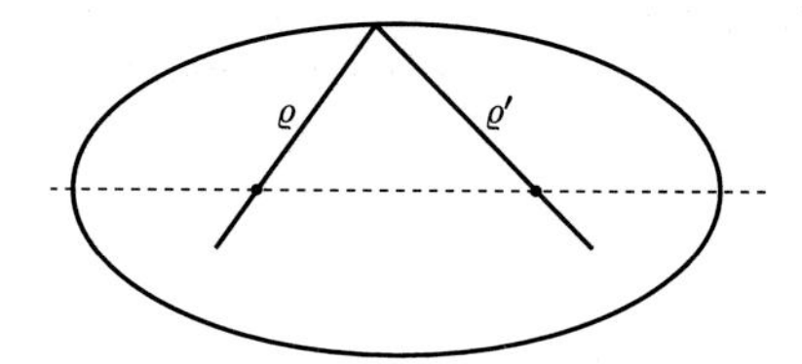

Figure 11.3: *Illustrating the Prolate Spheroid.* Show is a ray that passes through one focus, is reflected by the ellipse, then passes through the second focus. ρ and ρ' are the two distances from the point of incidence to the two foci.

which when cast in scalar form becomes,

$$\begin{cases} \xi' = \dfrac{1-\epsilon^2}{\mathcal{K}}\xi \\ \eta' = \dfrac{1-\epsilon^2}{\mathcal{K}}\eta \\ \zeta' = \dfrac{1}{\mathcal{K}}[(1+\epsilon^2)\zeta - 2\epsilon]. \end{cases} \tag{11.1}$$

Finally, also from Eq. 7.64, the distance from the point of incidence to the distal focus is

$$\varrho' = \varrho\frac{\mathcal{K}^2}{1-\epsilon^2}. \tag{7.64}$$

These relationships are shown in Fig. 11.3

11.3 Coupled Spheroids

The principle behind the modern schiefspiegler the coupling of two prolate spheroids are so that the distal focus of the first coincides with the proximal focus of the second. This we shall refer to as the *common focus*. Here ϵ_1 and r_1 are the eccentricity and semi latus rectum of the

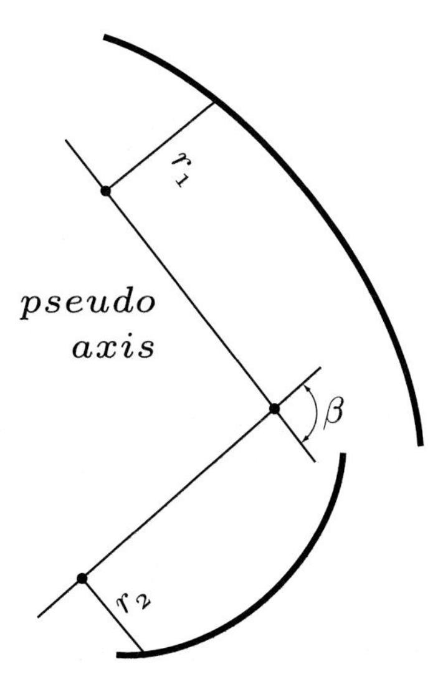

Figure 11.4: *Two Coupled Spheroids.* The distal focus of the first ellipsoid coincides with the proximal focus of the second. The two axes constitute the pseudo axis.

first spheroid and ϵ_2 and r_2 are those for the second. Let β be the angle between the axis of the first spheroid and that of the second. The system so obtained now has bilateral symmetry with the plane of symmetry determined by the two spheroid axes as shown in Fig. 11.4

Any ray through the proximal focus of the first spheroid, labeled **A** in Fig. 11.5, after reflection, must pass through the common focus, labeled **B**. Then, after a second reflection, through the distal focus of the second spheroid labeled **C**. Choose one such ray that lies entirely in the plane of symmetry and that makes an angle α with the axis of the first spheroid. This will be the pseudo axis; we will find a condition that will assure that it will be a line of symmetry for all other rays that, like it, pass though the first proximal focus and lie on the plane of symmetry.

In order to clarify what follows let $\vec{s}_1$ and $\vec{s}_2$ be unit vectors along the axes of the first and second spheroid, originating at the first **A** and second proximal foci **B**. Further, let $\vec{s}_0$ lie along the pseudo axis, originating at the proximal focus of the first spheroid **A**.

After reflection at the first spheroid the direction of the ray representing the pseudo axis will be designated as $\vec{s}_0{}'$; after reflection at the second spheroid its direction will be indicated by $\vec{s}_0{}''$.

Now take a more general ray through the first proximal focus with direction cosine vector $\mathbf{S} = (\xi,\ \eta,\ \zeta)$, the components of which are with respect to $\vec{s}_0$ the unit vector along the presumed pseudo axis. The y-component lies in the plane of symmetry and is normal to $\vec{s}_0$. Then this ray will have as direction cosines $\mathbf{S}_1 = (\xi_1,\ \eta_1,\ \zeta_1)$ where,

$$\begin{cases} \xi_1 = & \xi \\ \eta_1 = & \eta\cos\alpha + \zeta\sin\alpha \\ \zeta_1 = & -\,\eta\sin\alpha + \zeta\cos\alpha. \end{cases} \tag{11.2}$$

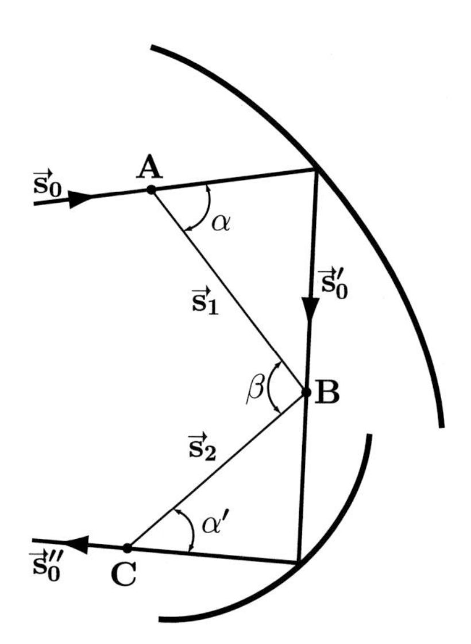

Figure 11.5: *Coupled Spheroids.* A is the location of the proximal focus of the first spheroid; C the distal focus of the second. B is where the distal focus of the first spheroid coincides with the proximal focus of the second. $\vec{s}_1$ and $\vec{s}_2$ are the spheroid axes; β is the subtended angle. The thicker line is the pseudo axis, α is the angle the pseudo axis makes with the axis of the first spheroid; α', is between the pseudo axis and the axis of the second spheroid in image space. $\vec{s}_0$, $\vec{s}_0{}'$ and $\vec{s}_0{}''$ are unit vectors along the pseudo axis in object space, in the space between the two reflections and in image space, respectively.

These components are relative to the original coordinate system whose z-axis, $\vec{s}_1$, is along the axis of the first spheroid. Again the y-axis lies in the plane of symmetry. Here we have done a coordinate rotation through the angle α at the point A.

Now we can invoke Eqs. 7.63 and 7.64 to find the point of incidence,

$$\mathbf{P} = \varrho_1(\xi_1,\ \eta_1\ ,\zeta_1), \tag{11.3}$$

where,

$$\varrho_1 = \frac{r_1}{1 - \epsilon_1\zeta_1}, \tag{11.4}$$

and the unit normal vector from Eq. 7.65,

$$\mathbf{N} = \frac{1}{\mathcal{K}_1}(\xi_1,\ \eta_1,\ \zeta_1 - \epsilon_1), \tag{11.5}$$

where, from Eq. 7.66 we get,

$$\mathcal{K}_1^2 = 1 + \epsilon_1^2 - 2\epsilon_1\zeta_1. \tag{11.6}$$

Referring to the first spheroid and Eq. 7.64, we get

$$\varrho_1' = \varrho_1 \frac{\mathcal{K}_1^2}{1 - \epsilon_1^2}, \tag{11.7}$$

the distance from the point of incidence to the common focus, B.

To Eq. 11.1 we apply the formulas for reflection given in Eq. 11.2 to get,

$$\begin{cases} \xi_1' = \dfrac{1-\epsilon_1^2}{\mathcal{K}_1}\xi \\ \eta_1' = \dfrac{1-\epsilon_1^2}{\mathcal{K}_1}(\eta\cos\alpha + \zeta\sin\alpha) \\ \zeta_1' = \dfrac{1}{\mathcal{K}_1}[(1+\epsilon_1^2)(-\eta\sin\alpha + \zeta\cos\alpha) - 2\epsilon_1]. \end{cases} \tag{11.8}$$

These components again need to be referred to $\vec{s}_1$ and to the plane of symmetry.

At the common focus we need to make another rotation of coordinates to a coordinate system whose z-axis, indicated by $\vec{s}_2$, is along the axis of the second spheroid. Recall that β is the angle between the spheroid axes. Then the coordinate rotation, just as in Eq. 11.2, is given by

$$\begin{cases} \xi_2 = \xi_1{}' \\ \eta_2 = \eta_1{}'\cos\beta + \zeta_1{}'\sin\beta \\ \zeta_2 = -\eta_1{}'\sin\beta + \zeta_1{}'\cos\beta, \end{cases} \tag{11.9}$$

where β is the angle between the two spheroid axes. Into Eq. 11.9 we substitute from Eq. 11.8 to get the components of $\mathbf{S}'$, now being referred to as $\vec{s}_2$. These are,

$$\begin{cases} \xi_2 = \dfrac{1-\epsilon_1^2}{\mathcal{K}_1}\xi \\ \eta_2 = \dfrac{1}{\mathcal{K}_1}[-2\epsilon_1\sin\beta + A\eta + B\zeta] \\ \zeta_2 = -\dfrac{1}{\mathcal{K}_1}[2\epsilon_1\cos\beta + C\eta + D\zeta], \end{cases} \tag{11.10}$$

where,

$$\begin{cases} A = (1-\epsilon_1^2)\cos\beta\cos\alpha - (1+\epsilon_1^2)\sin\beta\sin\alpha \\ B = (1-\epsilon_1^2)\cos\beta\sin\alpha + (1+\epsilon_1^2)\sin\beta\cos\alpha \\ C = (1-\epsilon_1^2)\sin\beta\cos\alpha + (1+\epsilon_1^2)\cos\beta\sin\alpha \\ D = (1-\epsilon_1^2)\sin\beta\sin\alpha - (1+\epsilon_1^2)\cos\beta\cos\alpha \end{cases} \tag{11.11}$$

at point B. All this is shown in Fig. 11.5.

We use again Eq. 7.66 to define,

$$\mathcal{K}_2^2 = 1 + \epsilon_2^2 - 2\epsilon_2\,\zeta_2 \tag{11.12}$$

and use it and Eq. 11.6 to get,

$$\begin{aligned}\mathcal{K}_1\mathcal{K}_2 &= (1+\epsilon_2^2)\mathcal{K}_1 + 2\epsilon_2(2\epsilon_1\cos\beta + C\eta + D\zeta)\\ &= (1+\epsilon_1^2)(1+\epsilon_2^2) + 4\epsilon_1\epsilon_2\cos\beta\\ &\quad +2\eta[\epsilon_1(1+\epsilon_2^2)\sin\alpha + \epsilon_2 C]\\ &\quad -2\zeta[\epsilon_1(1+\epsilon_2^2)\cos\alpha - \epsilon_2 D].\end{aligned} \tag{11.13}$$

By applying Eq. 7.64 we get the two relations

$$\begin{aligned}\varrho_2 &= \frac{r_2}{1-\epsilon_2\zeta_2} = \frac{r_2\mathcal{K}_2}{\mathcal{K}_2 + \epsilon_2(2\epsilon_1\cos\beta + C\eta + D\zeta)},\\ \varrho_2' &= \varrho_2\frac{\mathcal{K}_2}{1-\epsilon_2^2}.\end{aligned} \tag{11.14}$$

Finally we apply once again the reflection formulas in Eq. 11.1 to Eq. 11.10 and get $\mathbf{S}''$ whose components are,

$$\begin{cases}\xi_2' = \dfrac{(1-\epsilon_2^2)(1-\epsilon_1^2)}{\mathcal{K}_1\mathcal{K}_2}\xi\\[2mm] \eta_2' = \dfrac{1-\epsilon_2^2}{\mathcal{K}_1\mathcal{K}_2}(-2\epsilon_1\sin\beta + A\eta + B\zeta)\\[2mm] \zeta_2' = -\dfrac{1}{\mathcal{K}_1\mathcal{K}_2}[(1+\epsilon_2^2)(2\epsilon_1\cos\beta + C\eta + D\zeta) + 2\epsilon_2\mathcal{K}_1^2]\\[2mm] \quad = -\dfrac{1}{\mathcal{K}_1\mathcal{K}_2}(2P + Q\eta - R\zeta),\end{cases} \tag{11.15}$$

where,

$$\begin{cases}P = \epsilon_2(1+\epsilon_1^2) + \epsilon_1(1+\epsilon_2^2)\cos\beta\\ Q = 4\epsilon_1\epsilon_2\sin\alpha + (1+\epsilon_2^2)C\\ R = 4\epsilon_1\epsilon_2\cos\alpha - (1+\epsilon_2^2)D.\end{cases} \tag{11.16}$$

Equations 11.15 and 11.16 provides us with the direction cosines of this general ray at the distal focus of the second spheroid in terms of a coordinate system whose z-axis lies along $\vec{s}_2$, the axis of that second spheroid.

11.4 The Condition for the Pseudo Axis

We have tacitly assumed that the pseudo axis is that chief ray that makes an angle α with the axis of the first spheroid. It makes sense to call α' the angle that this same ray, in image space, makes with the axis of the second spheroid.

Now let $\xi = \eta = 0$ so that $\zeta = 1$. In what follows we signal this situation by a superior *bar* as in $\overline{\mathbf{S}}$. Then Eq. 11.13 provides the angle α' between the presumed pseudo axis and the axis of the second spheroid. These are given by,

$$\begin{cases} \overline{\mathcal{K}}_1\overline{\mathcal{K}}_2 \sin\alpha' = (1-\epsilon_2^2)(-2\epsilon_1\sin\beta + B) \\ \overline{\mathcal{K}}_1\overline{\mathcal{K}}_2 \cos\alpha' = -2P + R, \end{cases} \tag{11.17}$$

where, from Eq. 11.15,

$$\overline{\mathcal{K}}_1\overline{\mathcal{K}}_2 = (1+\epsilon_1^2)(1+\epsilon_2^2) + 4\epsilon_1\epsilon_2\cos\beta - 2[\epsilon_1(1+\epsilon_2^2)\cos\alpha - \epsilon_2 D]. \tag{11.18}$$

Next we take a ray lying in the plane of symmetry that makes an angle θ with the presumed pseudo axis so that $\xi = 0$ and $\eta = \sin\theta$, $\zeta = \cos\theta$. If θ' is the angle that this same ray makes with the pseudo axis in image space then, from Eq. 11.15, we get,

$$\begin{cases} \sin(\alpha'+\theta') = \dfrac{1-\epsilon_2^2}{\mathcal{K}_1\mathcal{K}_2}(-2\epsilon_1\sin\beta + A\sin\theta + B\cos\theta) \\ \cos(\alpha'+\theta') = -\dfrac{1}{\mathcal{K}_1\mathcal{K}_2}(2P + Q\sin\theta - R\cos\theta). \end{cases} \tag{11.19}$$

where, from Eq. 11.13,

$$\begin{aligned} \mathcal{K}_1\mathcal{K}_2 = {} & (1+\epsilon_1^2)(1+\epsilon_2^2) + 4\epsilon_1\epsilon_2\cos\beta \\ & +2[\epsilon_1(1+\epsilon_2^2)\sin\alpha + \epsilon_2 C]\sin\theta \\ & -2[\epsilon_1(1+\epsilon_2^2)\cos\alpha - \epsilon_2 D]\cos\theta. \end{aligned} \tag{11.20}$$

We use the identity

$$\sin\theta' = \sin(\alpha'+\theta')\cos\alpha' - \cos(\alpha'+\theta')\sin\alpha' \tag{11.21}$$

and Eqs. 11.17, 11.18 and 11.20 to get,

$$\begin{aligned} \sin\theta' = {} & \frac{1}{\mathcal{K}_1\mathcal{K}_2\overline{\mathcal{K}}_1\overline{\mathcal{K}}_2}\{2(BP - \epsilon_1 R\sin\beta)(1-\cos\theta) \\ & +[AR + BQ - 2(AP + \epsilon_1 Q\sin\beta)]\}. \end{aligned} \tag{11.22}$$

Recall that the condition for the pseudo axis is that it be a line of symmetry for all other chief rays lying in the plane of symmetry. Consider θ' as a function of θ. Then for this to hold $\theta'(-\theta) = -\theta'(\theta)$, so that θ' is an odd function of θ. Now look at the expression for the sine of θ' in Eq. 11.22. For it to be an odd function its numerator must be odd and its denominator must be even. For the numerator to be odd the coefficient of $1-\cos\theta$ must vanish. This gives us, with the aid of Eqs. 11.11 and 11.17, the following,

$$\begin{aligned} BP - \epsilon_1 R\sin\beta = {} & [(1-\epsilon_1^2)\cos\beta\sin\alpha + (1+\epsilon_1^2)\sin\beta\cos\alpha] \\ & \times[\epsilon_2(1+\epsilon_1^2) + \epsilon_1(1+\epsilon_2^2)\cos\beta] - \epsilon_1\sin\beta\{4\epsilon_1\epsilon_2\cos\alpha \\ & -(1+\epsilon_2^2)[(1-\epsilon_1^2)\sin\beta\sin\alpha - (1+\epsilon_1^2)\cos\beta\cos\alpha]\} \\ = {} & (1-\epsilon_1^2)\{[\epsilon_1(1+\epsilon_2^2) + \epsilon_2(1+\epsilon_1^2)\cos\beta]\sin\alpha \\ & +\epsilon_2(1-\epsilon_1^2)\sin\beta\cos\alpha\} - 0. \end{aligned} \tag{11.23}$$

Consider now the expression for the factor in the denominator, $\mathcal{K}_1\mathcal{K}_2$, found in Eq. 11.20. (Note that the product $\overline{\mathcal{K}}_1\overline{\mathcal{K}}_2$ does not enter into this calculation because it does not depend on θ.) For the coefficient of $\sin\theta$ to vanish

$$\epsilon_1(1+\epsilon_2^2)\sin\alpha+\epsilon_2 C=0, \tag{11.24}$$

which, on application of Eq. 11.11, yields,

$$\tan\alpha=-\frac{\epsilon_2(1-\epsilon_1^2)\sin\beta}{\epsilon_1(1+\epsilon_2^2)+\epsilon_2(1+\epsilon_1^2)\cos\beta}. \tag{11.25}$$

Note that the sum of the squares of numerator and denominator of the equation in Eq. 11.25 is,

$$\begin{aligned}
&\epsilon_2^2(1-\epsilon_1^2)^2\sin^2\beta+[\epsilon_1(1+\epsilon_2^2)+\epsilon_2(1+\epsilon_1^2)\cos\beta]^2\\
&\quad=\epsilon_2^2(1-\epsilon_1^2)^2(1-\cos^2\beta)+\epsilon_1^2(1+\epsilon_2^2)^2\\
&\qquad\qquad+2\epsilon_1\epsilon_2(1+\epsilon_1^2)(1+\epsilon_2^2)\cos\beta+\epsilon_2^2(1+\epsilon_1^2)\cos^2\beta\\
&\quad=\epsilon_1^2(1+\epsilon_2^2)^2+\epsilon_2^2(1-\epsilon_1^2)^2+2\epsilon_1\epsilon_2(1+\epsilon_1^2)(1+\epsilon_2^2)\cos\beta\\
&\qquad\qquad+\epsilon_2^2[(1+\epsilon_1^2)^2-(1-\epsilon_1^2)^2]\cos^2\beta\\
&\quad=(1+\epsilon_1^2\epsilon_2^2)(\epsilon_1^2+\epsilon_2^2)+2\epsilon_1\epsilon_2[(1+\epsilon_1^2\epsilon_2^2)+(\epsilon_1^2+\epsilon_2^2)]\cos\beta\\
&\qquad\qquad+4\epsilon_1^2\epsilon_2^2\cos^2\beta\\
&\quad=(1+\epsilon_1^2\epsilon_2^2+2\epsilon_1\epsilon_2\cos\beta)(\epsilon_1^2+\epsilon_2^2+2\epsilon_1\epsilon_2\cos\beta).
\end{aligned}$$

Now define,

$$\begin{cases}p^2=1+\epsilon_1^2\epsilon_2^2+2\epsilon_1\epsilon_2\cos\beta\\ q^2=\epsilon_1^2+\epsilon_2^2\;+2\epsilon_1\epsilon_2\cos\beta,\end{cases} \tag{11.26}$$

so that,

$$\begin{cases}\sin\alpha=-\epsilon_2(1-\epsilon_1^2)\sin\beta/pq\\ \cos\alpha=[\epsilon_1(1+\epsilon_2^2)+\epsilon_2(1+\epsilon_1^2)\cos\beta]/pq.\end{cases} \tag{11.27}$$

This assures that the denominator of the expression for $\sin\theta'$ will be an even function of θ. Substituting these expressions into Eq. 11.23 we see that that equation is satisfied also thus assuring that the numerator of $\sin\theta'$ is an odd function. It follows that $\sin\theta'$ is indeed an odd function and that Eq. 11.25 is indeed a condition for the system to possess a pseudo axis.

First of all note that, from Eq. 11.26,

$$\begin{cases}p^2+q^2=(1+\epsilon_1^2)(1+\epsilon_2^2)+4\epsilon_1\epsilon_2\cos\beta\\ p^2-q^2=(1-\epsilon_1^2)(1-\epsilon_2^2).\end{cases} \tag{11.28}$$

Next, referring to Eq. 11.20, it can be shown that,

$$\mathcal{K}_1\mathcal{K}_2=p^2+q^2-2pq\cos\theta, \tag{11.29}$$

so that,

$$\overline{\mathcal{K}}_1\overline{\mathcal{K}}_2 = (p-q)^2. \tag{11.30}$$

Now apply these results to Eq. 11.11 and get,

$$\begin{cases} A = (1-\epsilon_1^2)[\epsilon_2(1+\epsilon_1^2) + \epsilon_1(1+\epsilon_2^2)\cos\beta]/pq \\ B = \epsilon_1(p^2+q^2)\sin\beta/pq \\ C = \epsilon_1(1-\epsilon_1^2)(1+\epsilon_2^2)\sin\beta/pq \\ D = -[\epsilon_2(1-\epsilon_1^2)^2 + \epsilon_1(p^2+q^2)\cos\beta]/pq. \end{cases} \tag{11.31}$$

This and Eq. 11.17 yield up the following,

$$\begin{cases} \sin\alpha' = \epsilon_1(1-\epsilon_2^2)\sin\beta/pq \\ \cos\alpha' = [\epsilon_2(1+\epsilon_1^2) + \epsilon_1(1+\epsilon_2^2)\cos\beta]/pq. \end{cases} \tag{11.32}$$

The angle between the pseudo axis and the axis of the second ellipsoid is α'. The equations for α' given in Eq. 11.32 can be obtained from the equations for α in Eq. 11.27 by interchanging $'1'$ and $'2'$ and by changing the sign of β. This is tantamount to turning the system around.

11.5 Magnification and Distortion

The next step in this argument is the application of Eq. 11. 32 to Eq. 11.16 to get,

$$\begin{cases} P = pq\cos\alpha' \\ Q = (p^2-q^2)\sin\alpha' \\ R = (p^2+q^2)\cos\alpha'. \end{cases} \tag{11.33}$$

These, together with Eqs. 11.31 and 11.32, when applied to Eq. 11.15, yield the vector $\mathbf{S}''$, whose components are,

$$\begin{cases} \xi_2' = \frac{p^2-q^2}{p^2+q^2-2pq\zeta}\xi \\ \eta_2' = \frac{1}{p^2+q^2-2pq\zeta}\Big\{\big[\zeta(p^2+q^2)-2pq\big]\cos\alpha' \\ \qquad +\eta(p^2-q^2)\sin\alpha'\Big\} \\ \zeta_2' = \frac{1}{p^2+q^2-2pq\zeta}\Big\{\big[\zeta(p^2+q^2)-2pq\big]\sin\alpha' \\ \qquad -\eta(p^2-q^2)\cos\alpha'\Big\}, \end{cases} \tag{11.34}$$

which, as before, refer to $\vec{s}_2$, the vector along the axis of the second spheroid.

A final rotation of these vector components, through the angle α, as done in Eq. 11.2, gives us $\mathbf{S}'' = (\xi^*, \eta^*, \zeta^*)$, where,

$$\begin{cases} \xi^* = \dfrac{p^2 - q^2}{p^2 + q^2 - 2pq\zeta}\xi \\ \eta^* = \dfrac{p^2 - q^2}{p^2 + q^2 - 2pq\zeta}\eta \\ \zeta^* = \dfrac{\zeta(p^2 + q^2) - 2pq}{p^2 + q^2 - 2pq\zeta}. \end{cases} \tag{11.35}$$

These components are now referred to $\vec{s}_0{}''$, the unit vector along the pseudo axis in image space. This final rotation takes place at point C. See Fig. 11.5.

We can see from Eq. 11.35 that the pseudo axis, in image space, is indeed an axis of symmetry, not only for those chief rays lying in the plane of symmetry, but also for all chief rays. From this it follows that the attributes of chief rays that lie *in* the plane of symmetry will apply to *all* chief rays. So, with no loss of generality we find expressions for magnification and distortion for those rays lying *in* the plane of symmetry, confident that they will be valid for *all* chief rays.

So once again let $\xi = 0$ and let $\eta = \sin\theta$, $\zeta = \cos\theta$; then $\eta^* = \sin\theta'$ and $\zeta^* = \cos\theta'$. From Eq. 11.35 we can get,

$$\tan\theta' = \frac{(p^2 - q^2)\sin\theta}{(p^2 + q^2)\cos\theta - 2pq} = M(\theta)\tan\theta, \tag{11.36}$$

where,

$$M(\theta) = \frac{(p^2 - q^2)\cos\theta}{(p^2 + q^2)\cos\theta - 2pq}. \tag{11.37}$$

The magnification of the system must then be $M(0)$,

$$M(0) = M = \frac{p + q}{p - q}, \tag{11.38}$$

so that distortion is,

$$\begin{aligned} D(\theta) &= M(\theta) - M(0) \\ &= \frac{p + q}{p - q}\,\frac{2pq(1 - \cos\theta)}{(p^2 + q^2)\cos\theta - 2pq}. \end{aligned} \tag{11.39}$$

Distortion as a fraction of magnification then must be,

$$\overline{D}(\theta) = \frac{2pq(1 - \cos\theta)}{(p^2 + q^2)\cos\theta - 2pq}. \tag{11.40}$$

Now return to Eq. 11.38 and solve for p to get,

$$p = \frac{M + 1}{M - 1}\,q. \tag{11.41}$$

When this is substituted into Eq. 11.40 we get

$$\overline{D}(\theta) = \frac{(M^2 - 1)(1 - \cos\theta)}{(M^2 + 1)\cos\theta - (M^2 - 1)}, \tag{11.42}$$

which shows that distortion depends only on the system's angular magnification.

11.6 Conclusion

The modern schiefspiegler is an unusual system. The derivation of the condition for the existence of the pseudo axis reveals its unique properties. The good news, as shown in Eq. 11.35, is that its chief rays behave like chief rays in a rotationally symmetric system; that magnification is uniform in all directions. The bad news is that, once magnification is established, distortion is a fixed function of θ.

All of the results presented here can be extended to hyperboloids of two sheets by setting the value of the eccentricity (ϵ) to a value greater than unity or, in the case of telecentric systems, equal to unity. Moreover, by changing the sign of ϵ in Eq. 7.64, the roles of the proximal and distal foci are interchanged. However, all of our numerical work has been with prolate spheroids with the proximal and distal foci in their usual order.

A modest but comprehensive literature pertaining to the first-order properties of the systems without symmetry and their aberrations exists, beginning with a chapter in Kutter's book. Sands'[6] development of an aberration theory for what he calls plane symmetric systems and double-plane symmetric systems is certainly germane as is a more general work by Buchdahl.[7] The more modern and monumental work of Stone and Forbes,[8] which appeared in the later twentieth century, will certainly play an important role in the development of the schiefspiegler.

[6] Sands 1972, 1973a, and Sands 1973b.

[7] Buchdahl 1972.

[8] Stone and Forbes 1992a, Stone and Forbes 1992c, Stone and Forbes 1992d, and Stone and Forbes 1992e.

12 The Cartesian Oval and its Kin

Cartesian ovals are a class of refracting surfaces that image perfectly an object point into an image point. The first to find its algebraic formula was Descartes, though it is said that Kepler had obtained a numeric solution as early as 1602. Descartes was the first of a long line of investigators that had derived its formula independently, among whom was Maxwell; indeed, it is sometimes referred to as the Maxwellian oval.

It used to be that a new derivation was announced about every ten years giving us knowledgeable folk an opportunity to display our erudition,[1] but nowadays there is little interest in investigations of this kind.

There are two approaches to finding the equation for this kind of surface. One, which we will apply here, uses elementary algebra and relies on the fact that the optical path length, through any optical system, between two conjugate points, is constant. The other uses the refraction equation and the unit normal vector to the desired surface to find a partial differential equation whose solution is the appropriate formula.

The surface must be rotationally symmetric with the axis of symmetry being the line connecting the two conjugate points. Its equation is a quartic polynomial. However when either the object point or its image point is at *infinity* the equation degenerates into either a prolate spheroid or a hyperboloid of two sheets. When the surface is forced into the shape of a sphere, the two conjugate points become the *aplanatic points of the sphere* which we had already encountered in Chapter 9.

12.1 The Algebraic Method

Suppose a point on the required surface has the coordinates, $\mathbf{P} = (x,\ y,\ z)$, that the coordinates of the object point are $\mathbf{P_0} = (0,\ 0,\ -t) = -t\,\mathbf{Z}$ and that those of the image point are $\mathbf{P_1} = (0,\ 0,\ t_1) = t_1\mathbf{Z}$, where the z-axis is to be the line connecting $\mathbf{P_0}$ and $\mathbf{P_1}$. The unit vector $\mathbf{Z}$, of course, lies along the z-axis. Let the coordinate origin lie on this line at the point where it intersects the desired surface as is shown in Fig. 12.1. Then the distance d from the object point to $\mathbf{P}$ on the surface is,

$$d^2 = (\mathbf{P} + t\mathbf{Z})^2 = (x,\ y,\ z+t)^2 = \mathbf{P}^2 + 2tz + t^2, \tag{12.1}$$

and the distance d_1 from $\mathbf{P}$ to the image point is,

$$d_1^2 = (\mathbf{P} - t_1\mathbf{Z})^2 = (x,\ y,\ z-t_1)^2 = \mathbf{P}^2 - 2t_1 z + t_1^2, \tag{12.2}$$

[1] See, for example, O'Connell 1998.

The Mathematics of Geometrical and Physical Optics: The k-function and its Ramifications. O.N. Stavroudis

ISBN: 3-527-40448-1

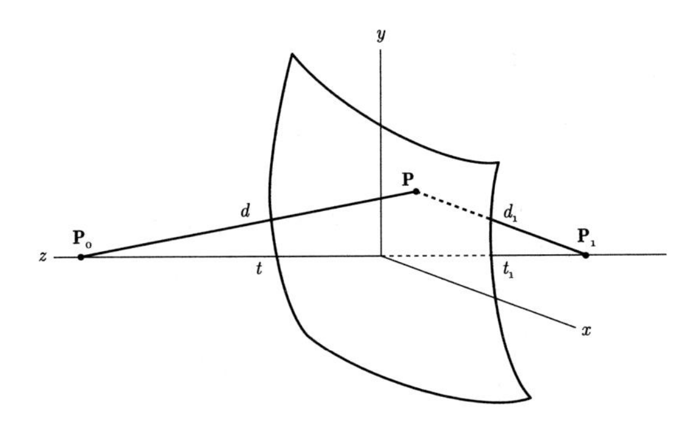

Figure 12.1: *The Cartesian Oval.* The object point is at $\mathbf{P}_0 = (0,\ 0,\ -t)$; its image, at $\mathbf{P}_1 = (0,\ 0,\ t_1)$; $\mathbf{P} = (x,\ y,\ z)$ is the point of incidence of a ray on the refracting surface.

Because $\mathbf{P_0}$ and $\mathbf{P_1}$ are to be conjugates the optical path length from $\mathbf{P_0}$ to $\mathbf{P}$ to $\mathbf{P_1}$ must be constant, so that

$$nd + n_1 d_1 = constant,$$

or

$$n\sqrt{\mathbf{P}^2 + 2tz + t^2} + n_1\sqrt{\mathbf{P}^2 - 2t_1 z + t_1^2} = constant. \tag{12.3}$$

If $\mathbf{P} = \mathbf{0}$ then Eq. 12.3 degenerates to

$$nt + n_1 t_1 = constant,$$

so that the equation for this surface must be

$$n\sqrt{\mathbf{P}^2 + 2tz + t^2} + n_1\sqrt{\mathbf{P}^2 - 2t_1 z + t_1^2} = nt + n_1 t_1. \tag{12.4}$$

We square both sides of this to get

$$\begin{aligned} n^2(\mathbf{P}^2 + 2tz) + n_1^2(\mathbf{P}^2 - 2t_1 z) - 2nn_1 tt_1 \\ = -2nn_1\sqrt{(\mathbf{P}^2 + 2tz) + t^2}\sqrt{(\mathbf{P}^2 - 2t_1 z) + t_1^2}. \end{aligned}$$

We again square both sides and again do further reductions to get

$$\begin{aligned} &\left[n^2(\mathbf{P}^2 + 2tz) - n_1^2(\mathbf{P}^2 - 2t_1 z)\right]^2 \\ &-4nn_1 tt_1\left[\mathbf{P}^2(n^2 + n_1^2) + 2z(n^2 t - n_1^2 t_1)\right] \\ &-4n^2 n_1^2\left[\mathbf{P}^2(t^2 + t_1^2) + 2tt_1 z(t_1 - t)\right] = 0. \end{aligned} \tag{12.5}$$

which can be simplified further:

$$\begin{aligned} &\left[(n^2 - n_1^2)\mathbf{P}^2 + 2z(n^2 t + n_1^2 t_1)\right]^2 \\ &-4nn_1(nt + n_1 t_1)\left[(nt_1 + n_1 t)\mathbf{P}^2 + 2tt_1 z(n - n_1)\right] = 0. \end{aligned} \tag{12.6}$$

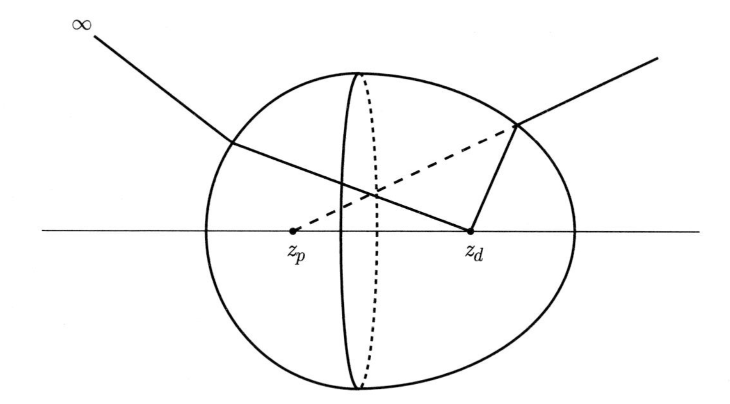

Figure 12.2: *The Prolate Spheroid.* The proximal focus is z_p; the distal focus, at z_d. Rays from an axial object point at ∞ converge to the distal focus.

This quartic polynomial is the equation of the ovaloid generated by rotating a Cartesian oval about its axis. An example is shown in Fig. 12.2.

The ovaloid is rotationally symmetric with the axis of symmetry being the line connecting the object and image points. Any ray connecting these two points determines a plane, the *plane of incidence.* Expressed as a scalar Eq. 12.6 becomes

$$\begin{aligned}&[(n^2-n_1^2)z^2+2z(n^2t+n_1^2t_1)+(n^2-n_1^2)(x^2+y^2)]^2\\&\quad-4nn_1(nt+n_1t_1)\big[(nt_1+n_1t)z^2+2tt_1z(n-n_1)\\&\quad+(nt_1+n_1t)(x^2+y^2)\big]=0.\end{aligned}\tag{12.7}$$

12.2 The Object at Infinity

Now we modify Eq. 12.7 by extending the object point to infinity. To do this we first divide Eq. 12.7 by t to get

$$\begin{aligned}&\left[\frac{1}{t}(n^2-n_1^2)z^2+2z\left(n^2+\frac{1}{t}n_1^2t_1\right)g+\frac{1}{t}(n^2-n_1^2)(x^2+y^2)\right]^2\\&\quad-4nn_1\left(n+\frac{1}{t}n_1t_1\right)\left[\left(\frac{1}{t}nt_1+n_1\right)(y^2+z^2)+2t_1(n-n_1)\right]=0.\end{aligned}\tag{12.8}$$

Next we let $t\to\infty$ so that Eq. 12.8 degenerates into

$$z^2(n^2-n_1^2)-2zn_1t_1(n-n_1)-n_1^2(x^2+y^2)=0$$

or

$$(n-n_1)[z^2(n+n_1)-2zn_1t_1]-n_1^2(x^2+y^2)=0.\tag{12.9}$$

This we multiply by $n + n_1$ and complete the square to get

$$(n - n_1)[z(n + n_1) - n_1t_1]^2 - n_1^2(x^2 + y^2)(n + n_1) = (n - n_1)n_1^2t_1^2, \tag{12.10}$$

a quadratic polynomial that clearly represents a conic section of revolution.

$$\frac{1}{B^2}\left(z - \frac{n_1t_1}{n + n_1}\right)^2 + \frac{1}{A^2}(x^2 + y^2) = 1, \tag{12.11}$$

where

$$\begin{cases} A^2 = \dfrac{n_1^2}{(n_1 + n)^2}t_1^2 \\ B^2 = \dfrac{n_1 - n}{n_1 + n}t_1^2. \end{cases} \tag{12.12}$$

12.3 The Prolate Spheroid

First suppose that $n_1 > n$. Then Eq. 12.11 can be interpreted as the equation of a prolate spheroid (an ellipsoid generated by revolving an ellipse about its major axis) with its center of symmetry at $z_c = A$ and whose eccentricity, from the standard formula[2] is

$$\epsilon = \sqrt{1 - \frac{B^2}{A^2}} = \sqrt{1 - \frac{n_1^2 - n^2}{n_1^2}} = \frac{n}{n_1} < 1. \tag{12.13}$$

By setting $x = y = 0$ in Eq. 12.11 we find the intercepts with the z-axis to be $z = 0$ and $z = 2a$, just as they should.

The distance from the center A to each of the two foci equals $\epsilon A = nt_1/(n_1 - n)$ so that their coordinates are

$$\begin{cases} z_p = A(1 - \epsilon) = \dfrac{n_1t_1}{n_1 + n}(1 - \dfrac{n}{n_1}) = t_1\dfrac{n_1 - n}{n_1 + n} \\ z_d = A(1 + \epsilon) = \dfrac{n_1t_1}{n_1 + n}(1 + \dfrac{n}{n_1}) = t_1, \end{cases} \tag{12.14}$$

where z_p is the *proximal focus* and z_d is the *distal focus*. This shows that the optical focus of the refracting prolate spheroid is at the spheroid's distal focus, the focus further from the vertex. Here the use of the word *focus* for two different concepts is indeed confusing.

Using a standard formula, we find the latus rectum to be,

$$\text{latus rectum} = 2\frac{B^2}{A} = 2t_1\frac{n_1 + n}{n_1}. \tag{12.15}$$

This is all shown in Fig. 12.3.

[2]The literature is indeed sparse. The Cartesian oval appears as an exercise in Buchdahl 1970, p. 310. Luneburg 1966, pp. 129–138 gives a detailed account (with part of which I disagree) and Stavroudis 1972, pp. 97–100 (in which I perpetuate Luneburg's incorrect methods). There is also Herzberger 1958, Chapter 17.

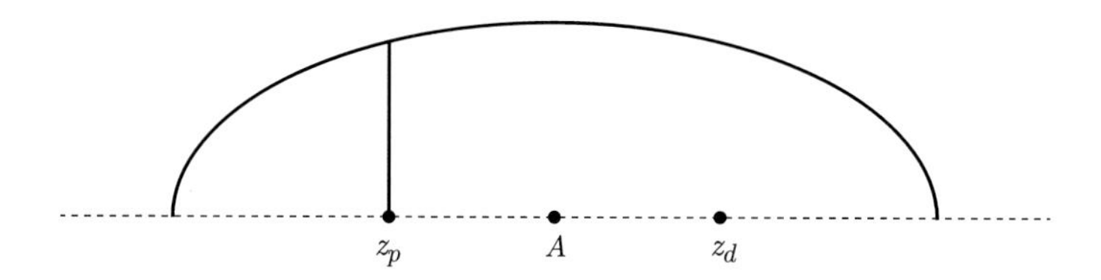

Figure 12.3: *Showing the ellipse.* z_p and z_d are the proximal and distal foci; A represents the center. The vertical line through z_p is the semi latus rectum.

12.4 The Hyperboloid of Two Sheets

Now suppose that $n > n_1$ so that Eq. 12.11 becomes,

$$-\frac{1}{B^2}\left(z - \frac{n_1 t_1}{n + n_1}\right)^2 + \frac{1}{A^2}(x^2 + y^2) = 1, \tag{12.16}$$

where,

$$\begin{cases} A^2 = \dfrac{n_1^2}{(n_1 + n)^2} t_1^2 \\ B^2 = \dfrac{n - n_1}{n_1 + n} t_1^2, \end{cases} \tag{12.17}$$

so the Cartesian oval degenerates into a hyperboloid of two sheets with its center of symmetry at $z_c = -A$ and with its z-intercepts at $z = 0$ and at $z = -2A$.

Again we find the eccentricity to be,

$$\epsilon = \sqrt{1 + \frac{B^2}{A^2}} = \sqrt{1 + \frac{n^2 - n_1^2}{n_1^2}} = \frac{n}{n_1} > 1. \tag{12.18}$$

As in the case of the spheroid, the two foci of the hyperboloid lie a distance ϵA on either side of the center of symmetry, $z_c = -A$ so that their coordinates are,

$$\begin{cases} z_p = -A(1 - \epsilon) = -\dfrac{n_1 t_1}{n + n_1}\left(1 - \dfrac{n}{n_1}\right) = t_1 \dfrac{n - n_1}{n + n_1} \\ z_d = -A(1 + \epsilon) = -\dfrac{n_1 t_1}{n + n_1}\left(1 + \dfrac{n}{n_1}\right) = -t_1. \end{cases} \tag{12.19}$$

This is analogous to the case of the prolate spheroid. To avoid unnecessary confusion we retain the notation for the proximal focus, z_p, the focus nearest the coordinate origin, and the distal focus, z_d, the more remote focus, which in this case has a negative z-coordinate. The distal focus, as in the case of the prolate spheroid, is the image of the infinite object point and is virtual. The latus rectum is the same as for the spheroid and is given by Eq. 12.15. This is all illustrated in Fig. 12.4.

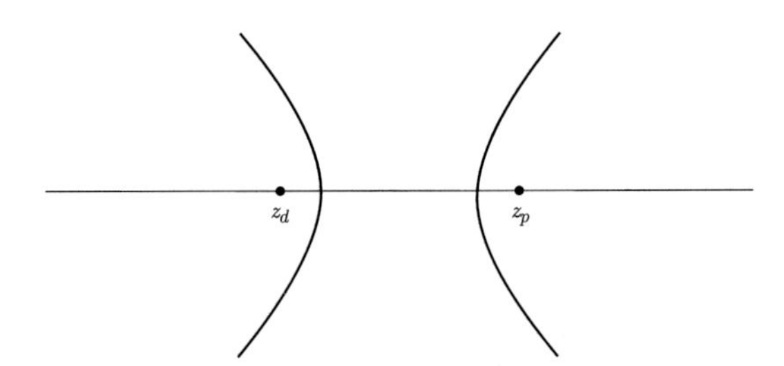

Figure 12.4: *The Hyperboloid of Two Sheets*. The proximal focus is z_p; the distal focus, at z_d the location of the virtual image of the axial object point at ∞.

12.5 Other Surfaces that Make Perfect Images

An interesting experiment relates to Eq. 12.6 which contains two factors each enclosed in brackets. Suppose we chose t and t_1 in such a way that these factors were equal and therefore the equation would split into the product of two quadratics. Let k be some factor yet to be determined and let

$$\begin{cases} n^2 + n_1^2 = k(n_1 t + n t_1) \\ n^2 t + n_1^2 t_1 = k t t_1 (n - n_1), \end{cases} \tag{12.20}$$

a pair of simultaneous equations that have two sets of solutions

$$t = -n_1/k, \qquad t_1 = nk \tag{12.21}$$

and

$$t = -(n+n_1)/k, \qquad t_1 = (n+n_1)/k. \tag{12.22}$$

Look at the first of these solutions, Eq. 12.21. First of all notice that $nt + n_1 t_1 = 0$ so that the optical path from object point to image point is *zero* and that the first term of Eq. 12.6 becomes

$$(n^2 - n_1^2)\mathbf{P}^2 - 2nn_1 z(n - n_1)/k = 0$$

which can be rearranged into

$$\left(z - \frac{nn_1}{k(n+n_1)}\right)^2 + x^2 + y^2 = \left(\frac{nn_1}{k(n+n_1)}\right)^2 \tag{12.23}$$

the equation of a sphere with z-intercepts at $z = 0$ and at $z = 2nn_1/k(n+n_1)$ and with a radius $r = nn_1/k(n+n_1)$ so that

$$k = \frac{nn_1}{r(n+n_1)}. \tag{12.24}$$

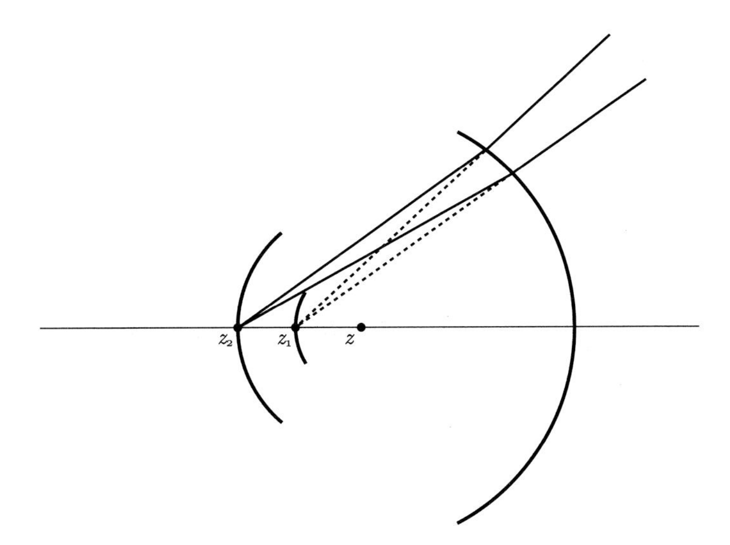

Figure 12.5: *Showing the Aplanatic Points and Surfaces of a sphere.* Rays focused on point z_1 in object space are refracted to focus on point z_2 in image space. z is the common center.

When this value of k is substituted back into Eq. 12.21 we get

$$t = -r(1 + n_1/n), \qquad t_1 = r(1 + n/n_1). \tag{12.25}$$

These represent distances to an object point and its image point that are imaged perfectly by the refracting sphere. These are the *aplanatic points of the sphere* and are shown in Fig. 12.5.

We can go further. Because the sphere is centrally symmetric two spheres can be generated by rotating the aplanatic points about the sphere's center. This generates a pair of spheres that are perfect images of one another. These are referred to as the *aplanatic surfaces* of a sphere. This is also illustrated in Fig. 12.5.

In days long past, before the advent of the computer, the aplanatic points and the aplanatic surfaces were used to trace rays graphically. Using only a straight edge and a compass, meridianal rays could be traced directly on the layout of a lens design, not with great accuracy but adequate for a quick and rough evaluation.

Now we return to Eq. 12.22 the other solution of the equations given in Eq. 12.20. As before we substitute these expressions into the equation for the Cartesian oval given in Eq. 12.6. The result is

$$\left(\mathbf{P}^2 - 2z\frac{n+n_1}{k}\right)\left[\left(\mathbf{P}^2 - 2z\frac{n+n_1}{k}\right) + 4\frac{nn_1}{k^2}\right] = 0. \tag{12.26}$$

We look at each quadratic factor separately. The left factor becomes

$$\left(z - \frac{n+n_1}{k}\right)^2 + x^2 + y^2 - \left(\frac{n+n_1}{k}\right)^2$$

another sphere that passes through the origin with radius $r = (n + n_1)/k$ so in this case

$$k = \frac{n + n_1}{r}. \tag{12.27}$$

When this is substituted into Eq. 12.22 we get

$$t = -r, \qquad t_1 = r \tag{12.28}$$

a statement of the unremarkable fact that the center of a sphere is its own perfect image.

The right hand factor of Eq. 12.26 is

$$\left(\mathbf{P}^2 - 2z\frac{n + n_1}{k}\right) + 4\frac{nn_1}{k^2} = 0$$

which becomes

$$x^2 + y^2 + (z - \frac{2n}{k})(z - \frac{2n_1}{k}) = 0, \tag{12.29}$$

the equation of an ovaloid with intercepts at $z = 2n/k$ and at $z = 2n_1/k$.

This concludes this shortest of chapters on the Cartesian oval. Unfortunately it seems to have no real applications except as a magnifying glass where the narrow end of the oval is cut off normal to the axis at the location of the image point and polished as a plane surface. Then when the plane surface is placed on a sheet of paper on which there is writing or some sort of diagram, a magnified virtual image is formed.

But it does give us a chance to display our erudition when the Cartesian oval is again rediscovered and republished.

13 The Pseudo Maxwell Equations

The first nine chapters are about Fermat's principle and its consequences. From this medieval relic we derived the eikonal equation and obtained its general solution in which the k-function first appeared as its arbitrary function. In Chapter 10 we parted from our preoccupation with geometrical optics and applied our previous results to the Maxwell equations for homogeneous, isotropic media. This lead us to a solution of the Maxwell equations in terms of the k-function.

We have demonstrated the power and the depth of Fermat's principle. In this chapter we speculate. Is there a more subtile relationship between the Maxwell equations and the axioms of geometrical optics? In this speculation we will return to the optics of inhomogeneous media and refer to our previous results in Chapters 1, 2 and 7. We will also write the Maxwell equations once again in terms of the arc length parameter s instead of the time parameter t.

13.1 Maxwell Equations for Inhomogeneous Media

A more general version than that used in Chapter 10 includes ϵ as a function of position while μ is assumed to be a constant.[1]

$$\begin{cases} \nabla \times \mathbf{E} = -\dfrac{1}{c}\dfrac{\partial \mathbf{B}}{\partial t} & \nabla \cdot \mathbf{B} = 0 & \mathbf{B} = \mu \mathbf{H} \\ \nabla \times \mathbf{H} = \dfrac{1}{c}\dfrac{\partial \mathbf{D}}{\partial t} & \nabla \cdot \mathbf{D} = 0 & \mathbf{D} = \epsilon \mathbf{E}. \end{cases} \tag{13.1}$$

However we need to make a change of variables from the time parameter t to the arc length parameter s as was done in Chapter 10

$$\text{(10.2)} \qquad \frac{\partial}{\partial t} = \frac{ds}{dt}\frac{\partial}{\partial s} = v\frac{\partial}{\partial s} = \frac{c}{n}\frac{\partial}{\partial s}.$$

The assumption that μ is constant leads to fatal inconsistencies; we must take it to be a function of position as is ϵ. To be consistent with reality we can assume that it is a slowly varying function of s.

$$\begin{cases} \nabla \times \mathbf{E} = -\sqrt{\dfrac{\mu}{\epsilon}}\dfrac{\partial \mathbf{H}}{\partial s} - \dfrac{\mu_s}{n}\mathbf{H} & \mu \nabla \cdot \mathbf{H} + \nabla \mu \cdot \mathbf{H} = 0 \\ \nabla \times \mathbf{H} = \sqrt{\dfrac{\epsilon}{\mu}}\dfrac{\partial \mathbf{E}}{\partial s} + \dfrac{\epsilon_s}{n}\mathbf{E} & \epsilon \nabla \cdot \mathbf{E} + \nabla \epsilon \cdot \mathbf{E} = 0 \end{cases} \tag{13.2}$$

[1] Borne and Wolf 1999, Chapter I.

The Mathematics of Geometrical and Physical Optics: The k-function and its Ramifications. O.N. Stavroudis

ISBN: 3-527-40448-1

13.2 The Frenet-Serret Equations

Now refer back to the vector trihedron in Chapter 2 and the relationship between $\mathbf{t}$, $\mathbf{n}$ and $\mathbf{b}$ and the Frenet-Serret equations expressed in terms of the directional derivative

$$\begin{aligned} \mathbf{t} &= \mathbf{n}\times\mathbf{b}, & \mathbf{t}_s &= (\mathbf{t}\cdot\nabla)\mathbf{t} = & &\frac{1}{\rho}\mathbf{n}, \\ \mathbf{n} &= \mathbf{b}\times\mathbf{t}, & \mathbf{n}_s &= (\mathbf{t}\cdot\nabla)\mathbf{n} = -\frac{1}{\rho}\mathbf{t} & &+\frac{1}{\tau}\mathbf{b}, \\ \mathbf{b} &= \mathbf{t}\times\mathbf{n}, & \mathbf{b}_s &= (\mathbf{t}\cdot\nabla)\mathbf{b} = & &-\frac{1}{\tau}\mathbf{n}. \end{aligned} \tag{2.12}$$

Let a train of wavefronts in an inhomogeneous medium be represented in the usual way by the vector function $\mathbf{W}(v,\ w,\ s)$. But the medium is not homogeneous so the results of Chapter 7 cannot apply. On the other hand the results of Chapter 1 do indeed apply and so we will make use of Eq. 1.29

(1.29 $$\frac{d}{ds}(n\mathbf{P}_s) = \nabla n,$$

and its equivalent expanded form

$$n\mathbf{P}_{ss}+(\nabla n\cdot\mathbf{P}_s)\mathbf{P}_s = \nabla n. \tag{1.30}$$

The results of Chapter 2 also apply and so we will make use of those forms of the ray equation, given in Eqs. 2.50 and 2.52, that involves the unit tangent vector $\mathbf{t}$ and the unit normal vector $\mathbf{n}$

$$\frac{n}{\rho}\mathbf{n}+(\nabla n\cdot\mathbf{t})\mathbf{t} = \nabla n. \tag{2.50}$$

and its equivalent

$$\frac{1}{\rho}\mathbf{n} = \frac{1}{n}\mathbf{t}\times(\nabla n\times\mathbf{t}), \tag{2.52}$$

as well as Eq. 2.51 that relates to the unit binormal vector $\mathbf{b}$,

$$\nabla n\cdot\mathbf{b} = 0. \tag{2.51}$$

We will also need

$$\nabla n\cdot\mathbf{n} = \frac{n}{\rho}, \tag{13.3}$$

also from Eq. 2.50.

13.3 Initial Calculations

These, Eqs. 2.50, 2.51 and 2.52, all relate to a ray, necessarily normal to the wavefront, as well as its vector trihedron, $\mathbf{t}$, $\mathbf{n}$ and $\mathbf{b}$. We have seen in Chapter 4 that the surface parameters can always be transformed according to a strict set of rules. Now we assume that this has already been done; that $\mathbf{W}_v$ is in the direction of $\mathbf{n}$ and that $\mathbf{W}_w$ is in the direction of $\mathbf{b}$. Through each point on the wavefront pass two of these lines of curvature that are perpendicular to each other. Moreover, since $\mathbf{N}$ is equal to $\mathbf{t}$ these parametric curves are geodesics as shown in Chapter 4.

The notation is indeed confusing. The ray's unit tangent vector $\mathbf{t}$ is also the unit normal vector $\mathbf{N}$ to the wavefront while $\mathbf{n}$ and $\mathbf{b}$ are normal to the ray and tangent to the wavefront surface. In an attempt to make things clearer we distinguish the two vector trihedrals associated with the two families of parametric curves by superscripts, $(^v)$ and $(^w)$; thus

$$\mathbf{t}^v = \mathbf{n}, \qquad \mathbf{n}^v = \mathbf{t}; \tag{13.4}$$

and

$$\mathbf{t}^w = \mathbf{b}, \qquad \mathbf{n}^w = \mathbf{t}; \tag{13.5}$$

so that

$$\begin{aligned} \mathbf{b}^v &= \mathbf{t}^v \times \mathbf{n}^v = \mathbf{n} \times \mathbf{t} = -\mathbf{b}, \\ \mathbf{b}^w &= \mathbf{t}^w \times \mathbf{n}^w = \mathbf{b} \times \mathbf{t} = \mathbf{n}. \end{aligned} \tag{13.6}$$

Recall that in Chapter 7 we had applied the Frenet-Serret equations to the $N,\ P,\ Q$ coordinate system to get the equations of generalized ray tracing. We do the same thing here to get, eventually, the expressions that I have called the *pseudo Maxwell equations*.

To each of these families of curves we next apply the Frenet-Serret equations, Eq. 2.12. For the v-curves we get

$$\begin{cases} (\mathbf{t}^v \cdot \nabla)\mathbf{t}^v = & \dfrac{1}{\rho_v}\mathbf{n}^v \\ (\mathbf{t}^v \cdot \nabla)\mathbf{n}^v = -\dfrac{1}{\rho_v}\mathbf{t}^v & & +\dfrac{1}{\tau_v}\mathbf{b}^v \\ (\mathbf{t}^v \cdot \nabla)\mathbf{b}^v = & -\dfrac{1}{\tau_v}\mathbf{n}^v. \end{cases} \tag{13.7}$$

When we apply Eqs. 13.4 and 13.5 to these we get

$$\begin{cases} (\mathbf{n} \cdot \nabla)\mathbf{n} \ = & \dfrac{1}{\rho_v}\mathbf{t} \\ (\mathbf{n} \cdot \nabla)\mathbf{t} \ = -\dfrac{1}{\rho_v}\mathbf{n} & & -\dfrac{1}{\tau_v}\mathbf{b} \\ -(\mathbf{n} \cdot \nabla)\mathbf{b} = & -\dfrac{1}{\tau_v}\mathbf{t}, \end{cases} \tag{13.8}$$

where $1/\rho_v$ is the curvature and $1/\tau_v$ the torsion of the family of v-curves. We do exactly the same thing for the w-curves to get

$$\begin{cases} (\mathbf{b}\cdot\nabla)\mathbf{b} = & \dfrac{1}{\rho_w}\mathbf{t} \\ (\mathbf{b}\cdot\nabla)\mathbf{t} = -\dfrac{1}{\rho_w}\mathbf{b} & + \dfrac{1}{\tau_w}\mathbf{n} \\ (\mathbf{b}\cdot\nabla)\mathbf{n} = & -\dfrac{1}{\tau_w}\mathbf{t}, \end{cases} \tag{13.9}$$

where, naturally, $1/\rho_w$ is the curvature and $1/\tau_w$ the torsion of each member of the family of w-curves.

Now refer back to Chapter 7 and the following equation

$$(7.34) \qquad \frac{1}{\sigma} = -\frac{1}{\tau_v} = \frac{1}{\tau_w}.$$

Although it had been applied to the expression for wavefronts in homogeneous, isotropic media its derivation was not limited to that case and can be applied to this situation. By applying Eq. 7.34 to Eqs. 13.8 and 13.9 we get

$$\begin{cases} (\mathbf{n}\cdot\nabla)\mathbf{n} = & \dfrac{1}{\rho_v}\mathbf{t} \\ (\mathbf{n}\cdot\nabla)\mathbf{t} = -\dfrac{1}{\rho_v}\mathbf{n} & + \dfrac{1}{\sigma}\mathbf{b} \\ (\mathbf{n}\cdot\nabla)\mathbf{b} = & -\dfrac{1}{\sigma}\mathbf{t}, \end{cases} \tag{13.10}$$

and

$$\begin{cases} (\mathbf{b}\cdot\nabla)\mathbf{b} = & \dfrac{1}{\rho_w}\mathbf{t} \\ (\mathbf{b}\cdot\nabla)\mathbf{t} = -\dfrac{1}{\rho_w}\mathbf{b} & + \dfrac{1}{\sigma}\mathbf{n} \\ (\mathbf{b}\cdot\nabla)\mathbf{n} = & -\dfrac{1}{\sigma}\mathbf{t}. \end{cases} \tag{13.11}$$

13.4 Divergence and Curl

In what follows we will need the following vector identities from the Appendix

$$(A.12) \qquad \nabla(\mathbf{E}\cdot\mathbf{F}) = (\mathbf{E}\cdot\nabla)\mathbf{F} + (\mathbf{F}\cdot\nabla)\mathbf{E} + \mathbf{E}\times(\nabla\times\mathbf{F}) + \mathbf{F}\times(\nabla\times\mathbf{E}),$$

$$(A.13) \qquad \nabla\cdot(\mathbf{E}\times\mathbf{F}) = \mathbf{F}\cdot(\nabla\times\mathbf{E}) - \mathbf{E}\cdot(\nabla\times\mathbf{F}),$$

$$(A.14) \qquad \nabla\times(\mathbf{E}\times\mathbf{F}) = (\mathbf{F}\cdot\nabla)\mathbf{E} - (\mathbf{E}\cdot\nabla)\mathbf{F} + \mathbf{E}(\nabla\cdot\mathbf{F}) - \mathbf{F}(\nabla\cdot\mathbf{E}).$$

In particular, we will need the fact that if $\mathbf{A}^2 = constant$ then, from Eq. A.7

$$(A.20) \qquad (\mathbf{A}\cdot\nabla)\mathbf{A} = -\mathbf{A}\times(\nabla\times\mathbf{A}).$$

Look first at the divergence of $\mathbf{t}$, in which we use repeatedly the vector products from Eqs. A.12 to A.14.

$$\begin{aligned}\nabla\cdot\mathbf{t} &= \nabla\cdot(\mathbf{n}\times\mathbf{b}) = \mathbf{b}\cdot(\nabla\times\mathbf{n}) - \mathbf{n}\cdot(\nabla\times\mathbf{b})\\ &= (\mathbf{t}\times\mathbf{n})\cdot(\nabla\times\mathbf{n}) - (\mathbf{b}\times\mathbf{t})\cdot(\nabla\times\mathbf{b})\\ &= \mathbf{t}\cdot[\mathbf{n}\times(\nabla\times\mathbf{n}) + \mathbf{b}\times(\nabla\times\mathbf{b})],\end{aligned}$$

to which we apply Eq. A.20, and get

$$\nabla\cdot\mathbf{t} = -\mathbf{t}\cdot[(\mathbf{n}\cdot\nabla)\mathbf{n} + (\mathbf{b}\cdot\nabla)\mathbf{b}],$$

which leads at once to, using Eqs. 13.10 and 13.11,

$$\nabla\cdot\mathbf{t} = -\mathbf{t}\cdot\left[\frac{1}{\rho_v}\mathbf{t} + \frac{1}{\rho_w}\mathbf{t}\right] = -\left(\frac{1}{\rho_v} + \frac{1}{\rho_w}\right), \tag{13.12}$$

a result obtained by Kline and Kay[2] using essentially the same methods. By using this same strategy we can also obtain

$$\begin{aligned}\nabla\cdot\mathbf{n} &= \nabla\cdot(\mathbf{b}\times\mathbf{t}) = \mathbf{t}\cdot(\nabla\times\mathbf{b}) - \mathbf{b}\times(\nabla\times\mathbf{t})\\ &= (\mathbf{n}\times\mathbf{b})\cdot(\nabla\times\mathbf{b}) - (\mathbf{t}\times\mathbf{n})\cdot(\nabla\times\mathbf{t})\\ &= \mathbf{n}\cdot[\mathbf{b}\times(\nabla\times\mathbf{b}) + \mathbf{t}\times(\nabla\times\mathbf{t})]\\ &= -\mathbf{n}\cdot[(\mathbf{b}\cdot\nabla)\mathbf{b} + (\mathbf{t}\cdot\nabla)\mathbf{t}]\\ &= -\mathbf{n}\cdot\left[\frac{1}{\rho_w}\mathbf{t} + \frac{1}{\rho}\mathbf{n}\right],\end{aligned}$$

resulting in

$$\nabla\cdot\mathbf{n} = -\frac{1}{\rho}. \tag{13.13}$$

Also

$$\begin{aligned}\nabla\cdot\mathbf{b} &= \nabla\cdot(\mathbf{t}\times\mathbf{n}) = \mathbf{n}\cdot(\nabla\times\mathbf{t}) - \mathbf{t}\times(\nabla\times\mathbf{n})\\ &= (\mathbf{b}\times\mathbf{t})\cdot(\nabla\times\mathbf{t}) - (\mathbf{n}\times\mathbf{b})\cdot(\nabla\times\mathbf{n})\\ &= \mathbf{b}\cdot[\mathbf{t}\times(\nabla\times\mathbf{t}) + \mathbf{n}\times(\nabla\times\mathbf{n})]\\ &= -\mathbf{b}\cdot[(\mathbf{t}\cdot\nabla)\mathbf{t} + (\mathbf{n}\cdot\nabla)\mathbf{n}]\\ &= -\mathbf{b}\cdot\left[\frac{1}{\rho}\mathbf{n} + \frac{1}{\rho_v}\mathbf{t}\right],\end{aligned}$$

so that

$$\nabla\cdot\mathbf{b} = 0. \tag{13.14}$$

[2] Kline and Kay 1965, pp. 184–186.

Next we return to the calculation of the curls of the three vectors that comprise the vector trihedron. In what follows we use Eqs. A.6, 7.34, 13.10 and 13.11 to get

$$\begin{aligned}\nabla \times \mathbf{t} &= \nabla \times (\mathbf{n} \times \mathbf{b}) \\ &= (\mathbf{b} \cdot \nabla)\mathbf{n} - (\mathbf{n} \cdot \nabla)\mathbf{b} + (\nabla \cdot \mathbf{b})\mathbf{n} - (\nabla \cdot \mathbf{n})\mathbf{b} \\ &= \tfrac{1}{\rho}\mathbf{b},\end{aligned} \tag{13.15}$$

$$\begin{aligned}\nabla \times \mathbf{n} &= \nabla \times (\mathbf{b} \times \mathbf{t}) \\ &= (\mathbf{t} \cdot \nabla)\mathbf{b} - (\mathbf{b} \cdot \nabla)\mathbf{t} + (\nabla \cdot \mathbf{t})\mathbf{b} - (\nabla \cdot \mathbf{b})\mathbf{t} \\ &= -\left(\tfrac{1}{\sigma} + \tfrac{1}{\tau}\right)\mathbf{n} - \tfrac{1}{\rho_v}\mathbf{b},\end{aligned} \tag{13.16}$$

and

$$\begin{aligned}\nabla \times \mathbf{b} &= \nabla \times (\mathbf{t} \times \mathbf{n}) \\ &= (\mathbf{n} \cdot \nabla)\mathbf{t} - (\mathbf{t} \cdot \nabla)\mathbf{n} + (\nabla \cdot \mathbf{n})\mathbf{t} - (\nabla \cdot \mathbf{t})\mathbf{n} \\ &= \tfrac{1}{\rho_w}\mathbf{n} + \left(\tfrac{1}{\sigma} - \tfrac{1}{\tau}\right)\mathbf{b}.\end{aligned} \tag{13.17}$$

13.5 Establishing the Relationship

We now write the electric field vector, $\mathbf{e}$, and the magnetic field vector, $\mathbf{h}$, as linear combinations of $\mathbf{n}$ and $\mathbf{b}$

$$\begin{cases}\mathbf{e} = \alpha\,\mathbf{n} + \beta\,\mathbf{b} \\ \mathbf{h} = -\beta\,\mathbf{n} + \alpha\,\mathbf{b},\end{cases} \tag{13.18}$$

so that $\mathbf{e} \cdot \mathbf{h} = \mathbf{0}$ and $\mathbf{e} \times \mathbf{h} = (\alpha^{\mathbf{2}} + \beta^{\mathbf{2}})\mathbf{t}$. We use the results of the previous section to calculate the curls, divergences and derivatives of $\mathbf{e}$ and $\mathbf{h}$ for substitution into Eqs. 13.2. First the divergences. From Eq. 13.2

$$\begin{aligned}&\epsilon\,\nabla \cdot \mathbf{e} + \nabla\epsilon \cdot \mathbf{e} = \mathbf{0} \\ &= \epsilon\big[\alpha(\nabla \cdot \mathbf{n}) + \beta(\nabla \cdot \mathbf{b}) + (\nabla\alpha \cdot \mathbf{n}) + (\nabla\beta \cdot \mathbf{b})\big] + \nabla\epsilon \cdot (\alpha\mathbf{n} + \beta\mathbf{b}) \\ &= -\alpha\epsilon/\rho + \epsilon(\mathbf{n} \cdot \nabla)\alpha + \epsilon(\mathbf{b} \cdot \nabla)\beta + \alpha(\mathbf{n} \cdot \nabla)\epsilon + \beta(\mathbf{b} \cdot \nabla)\mathbf{e},\end{aligned}$$

which leads to

$$\alpha\epsilon/\rho = (\mathbf{n} \cdot \nabla)(\epsilon\alpha) + (\mathbf{b} \cdot \nabla)(\epsilon\beta). \tag{13.19}$$

In exactly the same way, also from Eq. 13.2, we get the divergence of $\mathbf{h}$

$$\beta\mu/\rho = (\mathbf{n} \cdot \nabla)(\mu\beta) - (\mathbf{b} \cdot \nabla)(\mu\alpha). \tag{13.20}$$

The curls of $\mathbf{e}$ and $\mathbf{h}$ come next. Here we use Eqs. 13.16, 13.17 and 13.18 to get

$$\begin{aligned}
\nabla\times\mathbf{e} &= \nabla\times(\alpha\mathbf{n}+\beta\mathbf{b})\\
&= \alpha(\nabla\times\mathbf{n})+\beta(\nabla\times\mathbf{b})+(\nabla\alpha\times\mathbf{n})+(\nabla\beta\times\mathbf{b})\\
&= \alpha\left[-\left(\frac{1}{\sigma}+\frac{1}{\tau}\right)\mathbf{n}-\frac{1}{\rho_v}\mathbf{b}\right]+\frac{\beta}{\rho_w}\mathbf{n}+\left(\frac{1}{\sigma}+\frac{1}{\tau}\right)\mathbf{b}\\
&\quad +(\nabla\alpha\cdot\mathbf{t})\mathbf{b}-(\nabla\alpha\cdot\mathbf{b})\mathbf{t}+(\nabla\beta\cdot\mathbf{n})\mathbf{t}-(\nabla\beta\cdot\mathbf{t})\mathbf{n}\\
&= \mathbf{t}\left[-(\nabla\alpha\cdot\mathbf{b})+(\nabla\beta\cdot\mathbf{n})\right]+\mathbf{n}\left[-\alpha\left(\frac{1}{\sigma}+\frac{1}{\tau}\right)+\frac{\beta}{\rho_w}-\beta_s\right]\\
&\quad +\mathbf{b}\left[-\frac{\alpha}{\rho_v}+\beta\left(\frac{1}{\sigma}-\frac{1}{\tau}\right)+\alpha_s\right],
\end{aligned} \tag{13.21}$$

and

$$\begin{aligned}
\nabla\times\mathbf{h} &= \nabla\times(-\beta\mathbf{n}+\alpha\mathbf{b})\\
&= -\beta(\nabla\times\mathbf{n})+\alpha(\nabla\times\mathbf{b})-(\nabla\beta\times\mathbf{n})+(\nabla\alpha\times\mathbf{b})\\
&= -\beta\left[-\left(\frac{1}{\tau}+\frac{1}{\sigma}\right)\mathbf{n}-\frac{1}{\rho_v}\mathbf{b}\right]+\alpha\left[\frac{1}{\rho_w}\mathbf{n}+\left(\frac{1}{\sigma}-\frac{1}{\tau}\right)\mathbf{b}\right]\\
&\quad -\beta_s\mathbf{b}+(\nabla\beta\cdot\mathbf{b})\mathbf{t}+(\nabla\alpha\cdot\mathbf{n})\mathbf{t}-\alpha_s\mathbf{n}\\
&= \mathbf{t}\left[(\nabla\beta\cdot\mathbf{b})+(\nabla\alpha\cdot\mathbf{n})\right]+\mathbf{n}\left[\beta\left(\frac{1}{\sigma}+\frac{1}{\tau}\right)+\frac{\alpha}{\rho_w}-\alpha_s\right]\\
&\quad +\mathbf{b}\left[\frac{\beta}{\rho_v}+\alpha\left(\frac{1}{\sigma}-\frac{1}{\tau}\right)-\beta_s\right].
\end{aligned} \tag{13.22}$$

The derivatives are much simpler.

$$\begin{aligned}
\mathbf{e_s} &= (\alpha\,\mathbf{n}+\beta\,\mathbf{b})_s = \alpha\,\mathbf{n}_s+\beta\,\mathbf{b}_s+\alpha_s\,\mathbf{n}+\beta_s\,\mathbf{b}\\
&= \alpha\left(-\frac{1}{\rho}\mathbf{t}+\frac{1}{\tau}\mathbf{b}\right)+\beta\left(-\frac{1}{\tau}\mathbf{n}\right)+\alpha_s\,\mathbf{n}+\beta_s\,\mathbf{b}\\
&= -\frac{\alpha}{\rho}\mathbf{t}+\left(\alpha_s-\frac{\beta}{\tau}\right)\mathbf{n}+\left(\beta_s+\frac{\alpha}{\tau}\right)\mathbf{b},
\end{aligned} \tag{13.23}$$

and

$$\begin{aligned}
\mathbf{h_s} &= (-\beta\,\mathbf{n}+\alpha\,\mathbf{b})_s = -\beta\,\mathbf{n}_s+\alpha\,\mathbf{b}_s-\beta_s\,\mathbf{n}+\alpha_s\,\mathbf{b}\\
&= -\beta\left(-\frac{1}{\rho}\mathbf{t}+\frac{1}{\tau}\mathbf{b}\right)+\alpha\left(-\frac{1}{\tau}\mathbf{n}\right)-\beta_s\,\mathbf{n}+\alpha_s\,\mathbf{b}\\
&= \frac{\beta}{\rho}\mathbf{t}-\left(\beta_s+\frac{\alpha}{\tau}\right)\mathbf{n}+\left(\alpha_s-\frac{\beta}{\tau}\right)\mathbf{b}.
\end{aligned} \tag{13.24}$$

Eq. 13.19 becomes

$$\begin{aligned}
\alpha\epsilon/\rho &= (\mathbf{n}\cdot\nabla)(\alpha\epsilon)+(\mathbf{b}\cdot\nabla)(\beta\epsilon)\\
&= \epsilon[\nabla\alpha\cdot\mathbf{n}+\nabla\beta\cdot\mathbf{b}]+\alpha(\nabla\epsilon\cdot\mathbf{n})+\beta(\nabla\epsilon\cdot\mathbf{b})
\end{aligned} \tag{13.25}$$

and Eq. 13.20 turns into

$$\begin{aligned} \beta\mu/\rho &= (\mathbf{n}\cdot\nabla)(\beta\mu) + (\mathbf{b}\cdot\nabla)(\alpha\mu) \\ &= \mu[\nabla\beta\cdot\mathbf{n} - \nabla\alpha\cdot\mathbf{b}] + \beta(\nabla\mu\cdot\mathbf{n}) - \alpha(\nabla\mu\cdot\mathbf{b}). \end{aligned} \tag{13.26}$$

We substitute the expressions in Eqs. 13.21 and 13.24 back into Eqs. 13.2 to get

$$\begin{aligned} \nabla\times\mathbf{e} &= \mathbf{t}\left[-(\nabla\alpha\cdot\mathbf{b}) + (\nabla\beta\cdot\mathbf{n})\right] + \mathbf{n}\left[-\alpha\left(\frac{1}{\sigma}+\frac{1}{\tau}\right) + \frac{\beta}{\rho_w} - \beta_s\right] \\ &\quad +\mathbf{b}\left[-\frac{\alpha}{\rho_v} + \beta\left(\frac{1}{\sigma}-\frac{1}{\tau}\right) + \alpha_s\right] \\ &= -\sqrt{\frac{\mu}{\epsilon}}\left[\frac{\beta}{\rho}\mathbf{t} - \left(\beta_s + \frac{\alpha}{\tau}\right)\mathbf{n} + \left(\alpha_s - \frac{\beta}{\tau}\right)\mathbf{b}\right] - \frac{\mu_s}{n}(-\beta\mathbf{n} + \alpha\mathbf{b}), \end{aligned}$$

which leads to

$$\begin{aligned} &\mathbf{t}\left[-(\nabla\alpha\cdot\mathbf{b}) + (\nabla\beta\cdot\mathbf{n}) + \frac{\beta}{\rho}\sqrt{\frac{\mu}{\epsilon}}\right] \\ &+\mathbf{n}\left[-\alpha\left(\frac{1}{\sigma}+\frac{1}{\tau}\right) + \frac{\beta}{\rho_w} - \beta_s - \sqrt{\frac{\mu}{\epsilon}}\left(\beta_s + \frac{\alpha}{\tau}\right) - \beta\frac{\mu_s}{n}\right] \\ &+\mathbf{b}\left[\beta\left(\frac{1}{\sigma}-\frac{1}{\tau}\right) - \frac{\alpha}{\rho_v} + \alpha_s + \sqrt{\frac{\mu}{\epsilon}}\left(\alpha_s - \frac{\beta}{\tau}\right) + \alpha\frac{\mu_s}{n}\right] = 0. \end{aligned} \tag{13.27}$$

In exactly the same way we use Eqs. 13.22 and 13.23 to get

$$\begin{aligned} \nabla\times\mathbf{h} &= \mathbf{t}\left[(\nabla\alpha\cdot\mathbf{n}) + (\nabla\beta\cdot\mathbf{b})\right] + \mathbf{n}\left[\beta\left(\frac{1}{\sigma}+\frac{1}{\tau}\right) + \frac{\alpha}{\rho_w} - \alpha_s\right] \\ &\quad +\mathbf{b}\left[\frac{\beta}{\rho_v} + \alpha\left(\frac{1}{\sigma}-\frac{1}{\tau}\right) - \beta_s\right] \\ &= \sqrt{\frac{\epsilon}{\mu}}\left[-\frac{\alpha}{\rho}\mathbf{t} + \left(\alpha_s - \frac{\beta}{\tau}\right)\mathbf{n} + \left(\beta_s + \frac{\alpha}{\tau}\right)\mathbf{b}\right] + \frac{\epsilon_s}{n}(\alpha\mathbf{n} + \beta\mathbf{b}), \end{aligned}$$

which results in

$$\begin{aligned} &\mathbf{t}\left[\nabla\alpha\cdot\mathbf{n} + \nabla\beta\cdot\mathbf{b} + \frac{\alpha}{\rho}\sqrt{\frac{\epsilon}{\mu}}\right] \\ &+\mathbf{n}\left[\beta\left(\frac{1}{\sigma}+\frac{1}{\tau}\right) + \frac{\alpha}{\rho_w} - \alpha_s - \sqrt{\frac{\epsilon}{\mu}}\left(\alpha_s - \frac{\beta}{\tau}\right) - \alpha\frac{\epsilon_s}{n}\right] \\ &+\mathbf{b}\left[\alpha\left(\frac{1}{\sigma}-\frac{1}{\tau}\right) + \frac{\beta}{\rho_v} - \beta_s - \sqrt{\frac{\epsilon}{\mu}}\left(\beta_s + \frac{\alpha}{\tau}\right) - \beta\frac{\epsilon_s}{n}\right] = 0. \end{aligned} \tag{13.28}$$

From Eqs. 13.27 and 13.28 we get six scalar equations. Since each is a linear combination of the three orthogonal vectors $\mathbf{t}$, $\mathbf{n}$ and $\mathbf{b}$ set equal to *zero* the coefficients must also vanish resulting in

$$
\begin{aligned}
&a) \quad -(\nabla\alpha\cdot\mathbf{b})+(\nabla\beta\cdot\mathbf{n})+\frac{\beta}{\rho}\sqrt{\frac{\mu}{\epsilon}}=0,\\
&b) \quad -\frac{\alpha}{\sigma}+\frac{\beta}{\rho_w}-\left(\beta_s+\frac{\alpha}{\tau}\right)\left(1+\sqrt{\frac{\mu}{\epsilon}}\right)-\beta\frac{\mu_s}{n}=0,\\
&c) \quad \frac{\beta}{\sigma}-\frac{\alpha}{\rho_v}+\left(\alpha_s-\frac{\beta}{\tau}\right)\left(1+\sqrt{\frac{\mu}{\epsilon}}\right)+\alpha\frac{\mu_s}{n}=0,\\
&d) \quad \nabla\alpha\cdot\mathbf{n}+\nabla\beta\cdot\mathbf{b}+\frac{\alpha}{\rho}\sqrt{\frac{\epsilon}{\mu}}=0,\\
&e) \quad \frac{\beta}{\sigma}+\frac{\alpha}{\rho_w}-\left(\alpha_s-\frac{\beta}{\tau}\right)\left(1+\sqrt{\frac{\epsilon}{\mu}}\right)-\alpha\frac{\epsilon_s}{n}=0,\\
&f) \quad \frac{\alpha}{\sigma}+\frac{\beta}{\rho_v}-\left(\beta_s+\frac{\alpha}{\tau}\right)\left(1+\sqrt{\frac{\epsilon}{\mu}}\right)-\beta\frac{\epsilon_s}{n}=0.
\end{aligned}
\tag{13.29}
$$

Two more scalar equations come from Eqs. 13.25 and 13.26

$$
\begin{aligned}
&g) \quad \alpha\epsilon/\rho=\epsilon[\nabla\alpha\cdot\mathbf{n}+\nabla\beta\cdot\mathbf{b}]+\alpha(\nabla\epsilon\cdot\mathbf{n})+\beta(\nabla\epsilon\cdot\mathbf{b}),\\
&h) \quad \beta\mu/\rho=\mu[-\nabla\alpha\cdot\mathbf{b}+\nabla\beta\cdot\mathbf{n}]-\alpha(\nabla\mu\cdot\mathbf{b})+\beta(\nabla\mu\cdot\mathbf{n}).
\end{aligned}
\tag{13.30}
$$

Finally from Eqs. 2.56, 2.57 and 13.3 we get

$$
\begin{aligned}
\nabla n\cdot\mathbf{b}&=\frac{1}{2}\left[\sqrt{\frac{\mu}{\epsilon}}(\nabla\epsilon\cdot\mathbf{b})+\sqrt{\frac{\epsilon}{\mu}}(\nabla\mu\cdot\mathbf{b})\right]=0,\\
\nabla n\cdot\mathbf{n}&=\frac{1}{2}\left[\sqrt{\frac{\mu}{\epsilon}}(\nabla\epsilon\cdot\mathbf{n})+\sqrt{\frac{\epsilon}{\mu}}(\nabla\mu\cdot\mathbf{n})\right]=n/\rho,
\end{aligned}
$$

which leads to

$$
\begin{aligned}
&i) \quad \mu(\nabla\epsilon\cdot\mathbf{b})+\epsilon(\nabla\mu\cdot\mathbf{b})=0,\\
&j) \quad \mu(\nabla\epsilon\cdot\mathbf{n})+\epsilon(\nabla\mu\cdot\mathbf{n})=2\,n^2/\rho.
\end{aligned}
\tag{13.31}
$$

Now refer back to Eq. 13.29 and subtract e from c to get

$$
\alpha(\nabla\cdot\mathbf{t})+\frac{1}{n}(\sqrt{\epsilon}+\sqrt{\mu})^2\left(\alpha_s-\frac{\beta}{\tau}\right)+\frac{\alpha}{n}(\epsilon_s+\mu_s)=0,
\tag{13.32}
$$

as well as, by adding b and f,

$$
-\beta(\nabla\cdot\mathbf{t})-\frac{1}{n}(\sqrt{\epsilon}+\sqrt{\mu})^2\left(\beta_s+\frac{\alpha}{\tau}\right)-\frac{\beta}{n}(\epsilon_s+\mu_s)=0.
\tag{13.33}
$$

Now multiply Eq. 13.32 by β and Eq. 13.33 by α then add. The result after some reduction is

$$
\alpha_s\beta-\alpha\beta_s=\frac{1}{\tau}(\alpha^2+\beta^2).
\tag{13.34}
$$

Note that

$$\frac{\alpha_s\beta - \alpha\beta_s}{\alpha^2 + \beta^2} = \frac{d}{ds}\arctan\left(\frac{\alpha}{\beta}\right) = \frac{1}{\tau},$$

so that

$$\alpha = \beta \tan U \tag{13.35}$$

where

$$U = \int \frac{ds}{\tau}. \tag{13.36}$$

Next we substitute Eq. 13.35 into Eq. 13.32 and get

$$\tan U\left[\beta(\nabla\cdot\mathbf{t}) + \frac{\beta}{n}(\epsilon_s + \mu_s)\right] + \frac{1}{n}(\sqrt{\epsilon} + \sqrt{\mu})^2\left(\alpha_s - \frac{\beta}{\tau}\right) = 0 \tag{13.37}$$

and at the same time we make the same substitution into Eq. 13.33 resulting in

$$-\left[\beta(\nabla\cdot\mathbf{t}) + \frac{\beta}{n}(\epsilon_s + \mu_s)\right] - \frac{1}{n}(\sqrt{\epsilon} + \sqrt{\mu})^2\left(\beta_s - \frac{\beta}{\tau}\tan U\right) = 0. \tag{13.38}$$

By multiplying Eq. 13.38 by $\tan U$ and adding to Eq. 13.37 yields, after some reduction

$$\tau(\alpha_s - \beta_s \tan U) = \beta(1 + \tan^2 U)$$

or

$$\beta = \tau \cos U(\alpha_s \cos U - \beta_s \sin U) \tag{13.39}$$

14 The Perfect Lenses of Gauss and Maxwell

Like so many other myths in geometrical optics, a prefect lens is so rare as to be almost nonexistent. It should be rotationally symmetric and be able to image certain points in object space perfectly in image space. Examples of real prefect lenses are the Maxwell's fish eye, that was discussed in Chapters 1 and 2, and the sphere, in which the aplanatic surfaces are perfect conjugates, as was shown in Chapter 12.

14.1 Gauss' Approach

Gauss, in addition to setting the stage for paraxial optics, invented a system for analyzing the first order properties of a lens in the initial stages of its design. He began by defining six points, the *cardinal points*, with which we can trace rays graphically.

These cardinal points come in pairs. The two *foci*, one in object space, another in image space, are the images formed by the lens of an object point at infinity. The two *principal points* are conjugate points in object space and image space where both lie on the lens axis of symmetry. Any point placed on a plane perpendicular to the lens' axis that passes through the principal point in object space will be imaged on a similarly located plane in image space. The important thing here is that the magnification on these two planes is *unity*. The last two cardinal points are the *nodal points*, both lying on the lens axis with the property that any ray that passes through the object space nodal point must also pass through the nodal point in image space. Moreover, the two rays, one in object space and one in image space, must be parallel. Figure 14.1 is an attempt to show this. If the refractive index in object space and image space are equal, then the nodal points and the principal points coincide.

This kind of ray tracing had been used to establish the general structure of a lens at the beginning stages of an optical design. The y-$\overline{y}$ design system, used by some modern optical designers, is far superior.[1]

When the refractive indices of both object space and image space are air the nodal points and the principal points coincide. In this case the *focal length* of the lens equals the geometric distance between the focus and the *nodal/principal point*. This leads to an instrument, called the *nodal slide*, with which one can measure directly the focal length of a lens.

[1] Flores-Hernandez and Stavroudis 1996.

The Mathematics of Geometrical and Physical Optics: The k-function and its Ramifications. O.N. Stavroudis

ISBN: 3-527-40448-1

14.2 Maxwell's Approach

The mapping that Maxwell chose is the linear fractional transformation,

$$\begin{cases} x' = \dfrac{a_1x + b_1y + c_1z + d_1}{ax + by + cz + d} \\ y' = \dfrac{a_2x + b_2y + c_2z + d_2}{ax + by + cz + d} \\ z' = \dfrac{a_3x + b_3y + c_3z + d_3}{ax + by + cz + d}, \end{cases} \tag{14.1}$$

where $(x,\ y,\ z)$ represents a point in object space and where $(x',\ y',\ z')$ is its image. The inverse transform has an identical structure,

$$\begin{cases} x = \dfrac{A_1x' + B_1y' + C_1z' + D_1}{Ax' + By' + Cz' + D} \\ y = \dfrac{A_2x' + B_2y' + C_2z' + D_2}{Ax' + By' + Cz' + D} \\ z = \dfrac{A_3x' + B_3y' + C_3z' + D_3}{Ax' + By' + Cz' + D}, \end{cases} \tag{14.2}$$

another linear fractional transform. Here the coefficients, denoted by capital letters, are determinants whose elements are the coefficients that appear in Eq. 14.1.

The fractional-linear transformation maps planes into planes. This can be seen in the following way. Suppose a plane in object space is given by the equation,

$$px + qy + rz + s = 0, \tag{14.3}$$

into which we substitute $(x,\ y,\ z)$ from Eq. 14.2. The result is

$$\begin{aligned} &(pA_1 + qA_2 + rA_3 + sA)x' + (pB_1 + qB_2 + rB_3 + sB)y' \\ &+(pC_1 + qC_2 + rC_3 + sC)z' + (pD_1 + qD_2 + rD_3 + sD) = 0, \end{aligned} \tag{14.4}$$

clearly the equation of a plane in image space that is evidently the image of the plane in object space.

This transformation therefore maps planes into planes. Since a straight line can be represented as the intersection of two planes it follows that this transform maps straight lines into straight lines.[2]

From Eq. 14.1 we can see that the plane in object space, $ax + by + cz + d = 0$ is imaged at infinity in object space; from Eq. 14.2, infinity in object space is imaged into the plane $Ax' + By' + Cz' + D = 0$, in image space.

We have established coordinate systems in both object and image space. Now we impose conditions on the coefficients that bring the coordinate axes into correspondence. First we look

[2] Malacara and Thompson 2001, Chapter 1, pp. 1–8.

at a plane through the coordinate origin of object space perpendicular to the z-axis. Eq. 14.3, with $r = s = 0$, as its equation. From this and Eq. 14.4 we get the equation of its image,

$$(pA_1 + qA_2)x' + (pB_1 + qB_2)y' + (pC_1 + qC_2)z' + pD_1 + qD_2 = 0.$$

For this plane to pass through the image space coordinate origin and be perpendicular to the z'-axis, the coefficient of z' and the constant term must vanish identically, yielding,

$$C_1 = C_2 = D_1 = D_2 = 0. \tag{14.5}$$

Again using Eq. 14.3, by setting $q = 0$, we get the equation of a plane perpendicular to the y-axis whose image, from Eq. 14.4, is

$$(pA_1 + rA_3 + sA)x' + (pB_1 + rB_3 + sB)y' + (rC_3 + sC)x' + rD_3 + sD) = 0.$$

For this to be perpendicular to the y'-axis the coefficient of y' must equal zero, yielding,

$$B_1 = B_2 = B = 0, \tag{14.6}$$

The final step in this argument involves a plane perpendicular to the x-axis, obtained by setting $p = 0$ in Eq. 14.3. Its image, from Eq. 14.4, is

$$(qA_2 + rA_3 + sA)x' + (qB_2)y' + (rC_3 + sC)z' + rD_3 + sD = 0. \tag{14.7}$$

Now the coefficient of x' must vanish yielding the last of these conditions,

$$A_2 = A_3 = A = 0. \tag{14.8}$$

These conditions assure that the coordinate axes in image space are the images of those in object space. Nothing has been done to change any of the optical properties of this ideal instrument.

Substituting these, from Eqs. 14.5, 14.6, and 14.7, into Eq. 14.2 yields,

$$x = \frac{A_1 x'}{Cz' + D}, \quad y = \frac{B_2 y'}{Cz' + D}, \quad z = \frac{C_3 z' + D_3}{Cz' + D}. \tag{14.9}$$

It is a simple matter to invert this transformation to get,

$$\begin{aligned} A_1 &= \frac{cd_3 - c_3 d}{a_1}, \quad B_2 \quad = \frac{cd_3 - c_3 d}{b_2}, \\ C_3 &= -d, \quad C = c, \quad D_3 = d_3, \quad D = -c_3, \end{aligned} \tag{14.10}$$

so that Eqs. 14.1 and 14.2 now read,

$$x' = \frac{a_1 x}{cz + d}, \quad y' - \frac{b_2 y}{cz + d}, \quad z' - \frac{c_3 z - d_3}{cz + d}, \tag{14.11}$$

and

$$\begin{cases} x = \dfrac{x'(cd_3 - c_3 d)}{a_1(cz' - c_3)} \\ y = \dfrac{y'(cd_3 - c_3 d)}{b_2(cz' - c_3)} \\ z = -\dfrac{dz' - d_3}{cz' - c_3}. \end{cases} \tag{14.12}$$

Now we impose a restriction on the instrument itself by assuming that it is rotationally symmetric with the z-axis, and its image, the z'-axis as its axis of symmetry. Then in Eqs. 14.10 and 14.11, $b_2 = a_1$ so that we need only the y and z coordinates. They then degenerate into,

$$y' = \frac{a_1 y}{cz + d}, \quad z' = \frac{c_3 z + d_3}{cz + d}, \tag{14.13}$$

and

$$\begin{cases} y = \dfrac{y'(cd_3 - c_3 d)}{a_1(cz' - c_3)} \\ z = -\dfrac{dz' - d_3}{cz' - c_3}. \end{cases} \tag{14.14}$$

To recapitulate, the image of a point $(y,\ z)$ in object space is the point $(y',\ z')$ as determined by Eq. 14.12. Conversely, the image of $(y',\ z')$ in image space is the point $(y,\ z)$ in object space obtained from Eq. 14.13.

The plane perpendicular to the z-axis, given by the equation $cz + d = 0$, has, as its image, the plane at infinity as can be seen from Eq. 14.12. Therefore, $z_f = -d/c$ is the z-coordinate of the focal point of the instrument in object space. In exactly the same way we can find the z'-coordinate of the focal point in image space is $z'_f = c_3/c$, from Eq. 14.13. To summarize, we have shown that

$$\begin{cases} z_f = -d/c \\ z'_f = c_3/c. \end{cases} \tag{14.15}$$

At this point we make a change of variables shifting both coordinate origins to the two focal points thus,

$$\begin{cases} z = \bar{z} + z_f \\ z' = \bar{z}' + z'_f, \end{cases}$$

so that, from the second equation of Eq. 14.12, we get

$$\bar{z}' = \frac{cd_3 - c_3 d}{c^2 \bar{z}}. \tag{14.16}$$

From the second equation of 14.13 using the same transformation we obtain an identical result.

When the transformation is applied to the first equation of Eq. 14.12 we get

$$y' = \frac{a_1 y}{c\bar{z}}, \tag{14.17}$$

while the first equation of Eq. 14.13 yields

$$y = \frac{(cd_3 - c_3 d)y'}{a_1 c\bar{z}'}. \tag{14.18}$$

Now define *lateral magnification* as $m = y'/y$. Then from Eqs. 14.16 and 14.17 it follows that

$$m = \frac{a_1}{c\bar{z}} = \frac{a_1 c\bar{z}'}{cd_3 - cd_3}, \tag{14.19}$$

from which we can see that the conjugate planes of unit magnification are given by,

$$\begin{cases} \bar{z}_p = a_1/c \\ \bar{z}'_p = (cd_3 - cd_3)/a_1 c. \end{cases}$$

These are called the *principal planes* of the instrument. Now $\bar{z}_p$ and $\bar{z}'_p$ are the distances, along the axis of symmetry, between the foci and the principal points. These distances are called the *front* and *rear focal lengths* of the instrument and are denoted by f and f' respectively, thus

$$\begin{cases} f \ = a_1/c \\ f' = (cd_3 - c_3 d)/a_1 c. \end{cases}$$

Next we substitute these relations into Eq. 14.15 and get Newton's formula,

$$\bar{z}\bar{z}' = ff', \tag{14.20}$$

while from Eqs. 14.16 and 14.17 it follows that

$$\begin{cases} y' = fy/\bar{z} \\ y \ = f'y'/\bar{z}'. \end{cases} \tag{14.21}$$

Suppose now that y and $\bar{z}$ define a right triangle with a corner at the focus in object space and let θ be the angle subtended by the z-axis and the hypotenuse. Then the first equation of Eq. 14.21 becomes the familiar

$$y' = f\tan\theta. \tag{14.22}$$

Finally, let e equal the distance of an axial point in object space to the first principal point and let e' be the distance between its conjugate and the second principal point. Then it follows that

$$\begin{cases} e \ = \bar{z} - f \\ e' = \bar{z}' - f'. \end{cases} \tag{14.23}$$

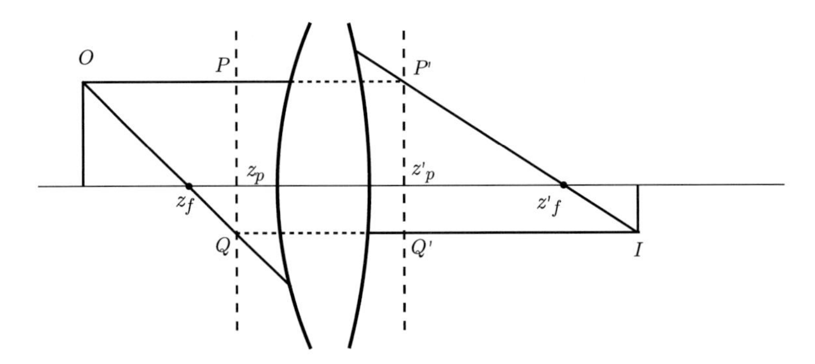

Figure 14.1: *Graphical construction of an object-image relationship.* The points z_f and z'_f are the instrument's foci and z_p and z'_p, its principal planes. From object point O, ray $\overline{OP}$ is parallel to the axis. Since P is on the object principal plane, its image P' must be at the same height. The image ray must therefore pass through $P'z'_f$. A second ray, $\overline{Oz_fQ}$, passes through the object focus and therefore must emerge in image space parallel to the axis. It must also pass through Q', the image of Q. The two rays cross at I, the image point.

Substituting these relations into Newton's formula, Eq. 14.20, results in the familiar

$$\frac{f}{e} + \frac{f'}{e'} = 1. \tag{14.24}$$

We have seen that straight lines are mapped into straight lines. Now we complete the argument and assume that such a line and its image constitute a single ray that is traced through the instrument.

From these results we can find object-image relationships using a graphic method. In Fig. 14.1 the points z_f and z'_f are the instrument's foci and z_p and z'_p its principal points. Let O be any object point. Let $\overline{OPP'}$ be a ray parallel to the axis, passing through P. Let its extension pass through P'. Since P and P' lie on the conjugate principal planes the ray in image space must pass through P'. Since this ray is parallel to the axis in object space its image must pass through z'_f. These two points determine completely this ray in image space. Now take a second ray, $\overline{Oz_fQ}$, through the object point O. Since it passes through z_f it must emerge in image space parallel to the axis. Since it passes through Q on the principal plane it must also pass through its image Q'. These two points determine this ray in image space. Where the two rays cross is I the image of O.

With this concept we can find a most important third pair of conjugates for which the instrument's *angular magnification* is unity. Then a ray passing through one of these points will emerge from the instrument and pass undeviated through the other. These are the *nodal* points.

Refer now to Fig. 14.2. Suppose a ray passes through the axis at z_0, at an angle θ, and intersects the principal plane at y_p. After passing through this ideal instrument it intersects the axis in image space at z'_0, at an angle θ', and passes through the principal plane at y'_p. Newton's formula, Eq. 14.20 provides a well-known relationship between z_0 and z'_0,

$$z_0 z'_0 = ff'. \tag{14.25}$$

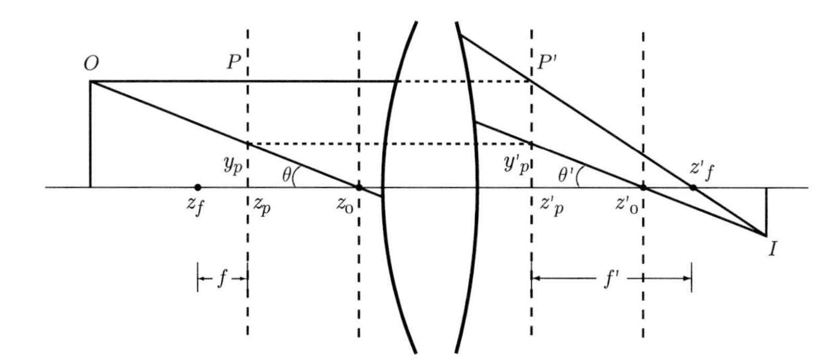

Figure 14.2: *The graphical construction of the nodal points.* The points z_f and z'_f are the two focal points and z_p and z'_p, the two principal planes. z_0 and z'_0 are the nodal points where $\theta = \theta'$. f and f' are the front and rear *focal lengths.*

Moreover, y_p and y'_p (which are not y_1 and y'_1) are equal since they represent the heights of conjugate points on the principle planes. From Fig. 14.2 we can see that

$$\begin{cases} y_p = -(z_0 - f)\tan\theta \\ y'_p = -(f' - z'_0)\tan\theta', \end{cases} \tag{14.26}$$

so that the angular magnification is given by

$$M = \frac{\tan\theta'}{\tan\theta} = \frac{z_0 - f}{f' - z'_0}.$$

With the aid of Eq. 14.25 this becomes

$$M = \frac{z_0}{f'}.$$

For z_0 and z'_0 to represent points where the angular magnification is unity it must be that,

$$\begin{cases} z_0 = f \\ z'_0 = f. \end{cases} \tag{14.27}$$

These nodal points are important for two reasons. In optical testing there is an instrument called the nodal slide which is based on the properties of the nodal points and is used to find, quickly and accurately, the focal length of a lens. A more subtle property is that images in image space bear the same perspective relationship to the second nodal point as do the corresponding objects in object space to the first nodal point.

With this in mind we make another change of variables; a translation of the z-axes to place the origins at the two nodal points. The new z-coordinates will be g and g'. The change is realized by,

$$\begin{cases} g = \bar{z} - \bar{z}_n = \bar{z} - f' \\ g' = \bar{z}' - \bar{z}'_n = \bar{z}' - f. \end{cases}$$

Again using Newton's formula, Eq. 14.20, we get

$$gg' + gf + g'f' = 0,$$

from which comes,

$$\frac{f}{g'} + \frac{f'}{g} + 1 = 0. \tag{14.28}$$

This concludes the study of the ideal optical instrument. We have found the six *cardinal points*, the foci, the principal points and the nodal points, solely from considering the Maxwell model of an ideal instrument. The model is a static one; there is no mention of wavefronts, velocities or refractive indices.

A Appendix. Vector Identities

We have already applied vectors to the parametric form of the Calculus of Variations in Chapter 1 where we also introduced the idea of a gradient vector in the context of total differentials. We elaborate that notation further here.

It is best to think of vectors as abstract objects with direction and length that can be represented as directed straight lines from some arbitrary but fixed coordinate origin to a point in space whose coordinates are, say, $(x,\ y,\ z)$. We will indicate a vector by a bold faced capital letter, thus; $\mathbf{P} = (x,\ y,\ z)$. From the Pythagorean theorem its length is $|\mathbf{P}| = \sqrt{x^2 + y^2 + z^2}$. Then the unit vector in the direction of $\mathbf{P}$, often referred to as the *direction cosine vector*, is $\mathbf{P}/\sqrt{x^2 + y^2 + z^2} = \mathbf{P}/|\mathbf{P}|$.

Let $\mathbf{Q} = (x_1,\ y_1,\ z_1)$ be a second vector. The *scalar product* or *dot product* of the two vectors is defined as $\mathbf{P} \cdot \mathbf{Q} = xx_1 + yy_1 + zz_1$. We also use the notation $\mathbf{P}^2 = \mathbf{P} \cdot \mathbf{P}$ so that we can write the vector's length as $|\mathbf{P}| = \sqrt{\mathbf{P}^2}$.

The addition of these two vectors is a third vector $\mathbf{R} = \mathbf{P} + \mathbf{Q}$ which is obtained by translating $\mathbf{Q}$ parallel to itself and attaching it to the end of $\mathbf{P}$. This is shown in Fig. A.1. The square of the length of $\mathbf{R}$ then must be

$$\mathbf{R}^2 = (\mathbf{P} + \mathbf{Q})^2 = \mathbf{P}^2 + \mathbf{Q}^2 + 2\,\mathbf{P} \cdot \mathbf{Q}. \tag{A.1}$$

Let α be the angle between $\mathbf{P}$ and $\mathbf{Q}$ and let β its supplement so that $\alpha + \beta = 180°$.

Then by the law of cosines

$$\mathbf{R}^2 = \mathbf{P}^2 + \mathbf{Q}^2 - 2\sqrt{\mathbf{P}^2}\,\sqrt{\mathbf{Q}^2}\,\cos\beta, \tag{A.2}$$

so that

$$\mathbf{P} \cdot \mathbf{Q} = -\sqrt{\mathbf{P}^2}\,\sqrt{\mathbf{Q}^2}\,\cos\beta = \sqrt{\mathbf{P}^2}\,\sqrt{\mathbf{Q}^2}\,\cos\alpha. \tag{A.3}$$

Thus the scalar product of two vectors is equal to the product of their lengths times the cosine of the subtended angle. It follows that if $\mathbf{P}$ and $\mathbf{Q}$ are perpendicular then $\mathbf{P} \cdot \mathbf{Q} = 0$.

Next, suppose a third vector $\mathbf{R} = (p,\ q,\ r)$ is perpendicular to both $\mathbf{P}$ and $\mathbf{Q}$. Then $\mathbf{R} \cdot \mathbf{P} = 0$ and $\mathbf{R} \cdot \mathbf{Q} = 0$, or in scalar form

$$\begin{cases} px + qy + rz &= 0 \\ px_1 + qy_1 + rz_1 &= 0, \end{cases} \tag{A.4}$$

which when solved for $\mathbf{R}$ yields

$$\mathbf{R} = (p,\ q,\ r) = (y\,z_1 - z\,y_1,\ -x\,z_1 + z\,x_1,\ x\,y_1 - y\,x_1). \tag{A.5}$$

The Mathematics of Geometrical and Physical Optics: The k-function and its Ramifications. O.N. Stavroudis

ISBN: 3-527-40448-1

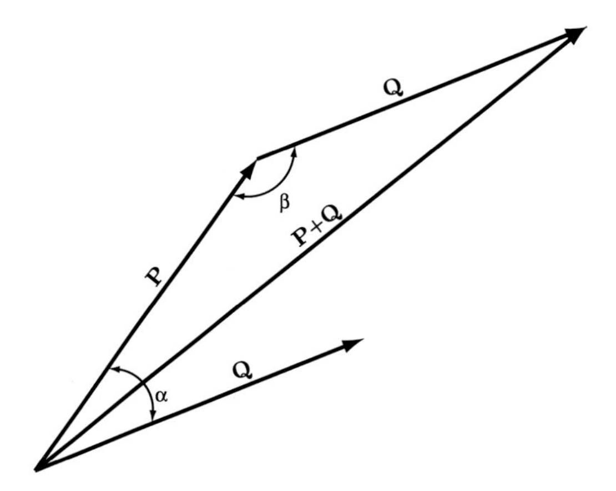

Figure A.1: *Addition of Two Vectors.*

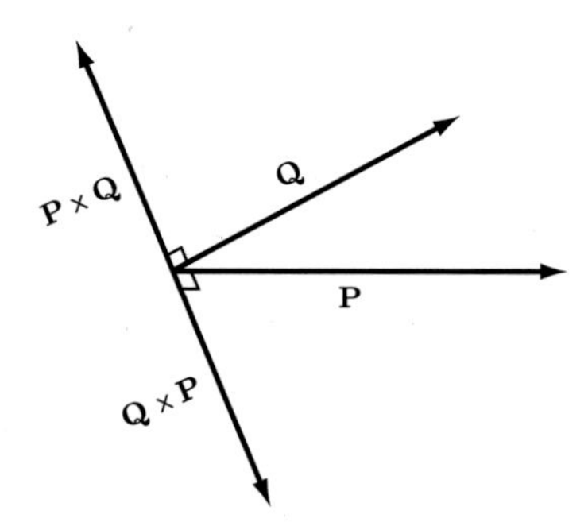

Figure A.2: *Vector Product.*

This is the definition of the *vector product* or *cross product* of the two vectors $\mathbf{P}$ and $\mathbf{Q}$ which we will write as $\mathbf{R} = \mathbf{P} \times \mathbf{Q}$.

Using Eq. A.5 and the vector identity in Eq. A.3 we can get

$$\mathbf{R}^2 = (\mathbf{P} \times \mathbf{Q})^2 = \mathbf{P}^2\mathbf{Q}^2 - (\mathbf{P} \cdot \mathbf{Q})^2 = \mathbf{P}^2\mathbf{Q}^2(1 - \cos^2 \alpha) = \mathbf{P}^2\mathbf{Q}^2 \sin^2 \alpha,$$

where the angle subtending $\mathbf{P}$ and $\mathbf{Q}$ as above is α.

From this we can see that the vector product of two parallel vectors must vanish. It is clear from Eq. A.5 that $\mathbf{P} \times \mathbf{Q} = -\mathbf{Q} \times \mathbf{P}$ showing that the vector product does not commute. Figure A.2 is an attempt to illustrate these relationships.

A.1 Algebraic Identities

These involve the manipulation of the vector components,

$$\mathbf{A} \times (\mathbf{B} \times \mathbf{C}) = (\mathbf{A} \cdot \mathbf{C})\mathbf{B} - (\mathbf{A} \cdot \mathbf{B})\mathbf{C} \tag{A.6}$$

$$(\mathbf{A} \times \mathbf{B}) \cdot (\mathbf{C} \times \mathbf{D}) = (\mathbf{A} \cdot \mathbf{C})(\mathbf{B} \cdot \mathbf{D}) - (\mathbf{A} \cdot \mathbf{D})(\mathbf{B} \cdot \mathbf{C}) \tag{A.7}$$

From Eq. A.6 we can get,

$$(\mathbf{A} \times \mathbf{B}) \times (\mathbf{C} \times \mathbf{D}) = (\mathbf{A} \times \mathbf{B}) \cdot \mathbf{D}\,\mathbf{C} - (\mathbf{A} \times \mathbf{B}) \cdot \mathbf{C}\,\mathbf{B} \tag{A.8}$$

and from Eq. A.7 comes,

$$(\mathbf{A} \times \mathbf{B})^2 = \mathbf{A}^2\,\mathbf{B}^2 - (\mathbf{A} \cdot \mathbf{B})^2. \tag{A.9}$$

A.2 Identities Involving First Derivatives

$$\nabla \cdot (\gamma\,\mathbf{E}) = \nabla\gamma \cdot \mathbf{E} + \gamma\,\nabla \cdot \mathbf{E}. \tag{A.10}$$

$$\nabla \times (\gamma\,\mathbf{E}) = \nabla\gamma \times \mathbf{E} + \gamma\,\nabla \times \mathbf{E}. \tag{A.11}$$

$$\nabla(\mathbf{E} \cdot \mathbf{F}) = (\mathbf{E} \cdot \nabla)\mathbf{F} + (\mathbf{F} \cdot \nabla)\mathbf{E} + \mathbf{E} \times (\nabla \times \mathbf{F}) + \mathbf{F} \times (\nabla \times \mathbf{E}), \tag{A.12}$$

$$\nabla \cdot (\mathbf{E} \times \mathbf{F}) = \mathbf{F} \cdot (\nabla \times \mathbf{E}) - \mathbf{E} \cdot (\nabla \times \mathbf{F}), \tag{A.13}$$

$$\nabla \times (\mathbf{E} \times \mathbf{F}) = (\mathbf{F} \cdot \nabla)\mathbf{E} - (\mathbf{E} \cdot \nabla)\mathbf{F} + \mathbf{E}(\nabla \cdot \mathbf{F}) - \mathbf{F}(\nabla \cdot \mathbf{E}). \tag{A.14}$$

Note the directional derivatives in Eqs. A.12 and A.14.

A.3 Identities Involving Second Derivatives

Here we introduce the Lagrangian operator,

$$\nabla^2 = \partial^2/\partial x^2 + \partial^2/\partial y^2 + \partial^2/\partial z^2. \tag{A.15}$$

$$\nabla \cdot (\nabla\Phi) = \nabla^2\Phi, \tag{A.16}$$

$$\nabla(\nabla \cdot \mathbf{A}) = \nabla^2\mathbf{A} - \nabla \times (\nabla \times \mathbf{A}), \tag{A.17}$$

$$\nabla \times (\nabla\Phi) = 0, \tag{A.18}$$

$$\nabla \cdot (\nabla \times \mathbf{A}) = 0. \tag{A.19}$$

From Eq. A.17, if $\mathbf{A}^2 = constant$ then,

$$(\mathbf{A} \cdot \nabla)\mathbf{A} = -\mathbf{A} \times (\nabla \times \mathbf{A}). \tag{A.20}$$

A.4 Gradient

In Cartesian terms the gradient of a function $f(x,y,z)$ is defined as $\nabla f = (\partial f/\partial x, \partial f/\partial y, \partial f/\partial z)$. To switch to the system of generalized coordinates we assume that each of the arguments of f is a component of **W** and therefore a function of $u,\ v$ and w. We take the derivatives of f with respect to these generalized coordinates and use the chain rule to get

$$\frac{\partial f}{\partial v} = \frac{\partial f}{\partial x}x_v + \frac{\partial f}{\partial y}y_v + \frac{\partial f}{\partial z}z_v$$
$$\frac{\partial f}{\partial w} = \frac{\partial f}{\partial x}x_w + \frac{\partial f}{\partial y}y_w + \frac{\partial f}{\partial z}z_w$$
$$\frac{\partial f}{\partial s} = \frac{\partial f}{\partial x}x_s + \frac{\partial f}{\partial y}y_s + \frac{\partial f}{\partial z}z_s$$

or in matrix form

$$\begin{pmatrix} x_v\ y_v\ z_v \\ x_w\ y_w\ z_w \\ x_s\ y_s\ z_s \end{pmatrix} \begin{pmatrix} f_x \\ f_y \\ f_z \end{pmatrix} = \begin{pmatrix} f_v \\ f_w \\ f_s \end{pmatrix}.$$

The determinant of coefficients is

$$\begin{vmatrix} x_v\ y_v\ z_v \\ x_w\ y_w\ z_w \\ x_s\ y_s\ z_s \end{vmatrix} = x_s(y_v z_w - y_w z_v) - y_s(x_v z_w - x_w z_v) + z_s(x_v y_w - y_v z_w)$$
$$= \mathbf{W}_s \cdot (\mathbf{W}_v \times \mathbf{W}_w) = D,$$

where D is defined in Eqs. 4.8 and 6.10.

With Cramer's rule we solve for $f_x,\ f_y$, and f_z to get

$$Df_x = \begin{vmatrix} f_v\ y_v\ z_v \\ f_w\ y_w\ z_w \\ f_s\ y_s\ z_s \end{vmatrix} = f_v \begin{vmatrix} y_w\ y_s \\ z_s\ z_w \end{vmatrix} - f_w \begin{vmatrix} y_v\ y_s \\ z_s\ z_v \end{vmatrix} + f_s \begin{vmatrix} y_v\ y_w \\ z_w\ z_v \end{vmatrix}$$

$$Df_y = \begin{vmatrix} x_v\ f_v\ z_v \\ z_s\ x_w\ f_w \\ x_s\ f_s\ z_s \end{vmatrix} = -f_v \begin{vmatrix} x_w\ x_s \\ z_s\ z_w \end{vmatrix} + f_w \begin{vmatrix} x_v\ x_s \\ z_s\ z_v \end{vmatrix} - f_s \begin{vmatrix} x_v\ x_w \\ z_w\ z_v \end{vmatrix}$$

$$Df_z = \begin{vmatrix} x_v\ y_v\ f_v \\ x_w\ y_w\ f_w \\ x_s\ y_s\ f_s \end{vmatrix} = f_v \begin{vmatrix} x_w\ x_s \\ y_s\ y_w \end{vmatrix} - f_w \begin{vmatrix} x_v\ x_s \\ y_s\ y_v \end{vmatrix} + f_s \begin{vmatrix} x_v\ x_w \\ y_w\ y_v \end{vmatrix}.$$

Then the gradient of f can be written as

$$
\begin{aligned}
D\nabla f &= D(f_x,\ f_y,\ f_z)\\
&= \left(f_v \begin{vmatrix} y_w & y_s \\ z_s & z_w \end{vmatrix} - f_w \begin{vmatrix} y_v & y_s \\ z_s & z_v \end{vmatrix} + f_s \begin{vmatrix} y_v & y_w \\ z_w & z_v \end{vmatrix}, \right.\\
&\quad -f_v \begin{vmatrix} x_w & x_s \\ z_s & z_w \end{vmatrix} + f_w \begin{vmatrix} x_v & x_s \\ z_s & z_v \end{vmatrix} - f_s \begin{vmatrix} x_v & x_w \\ z_w & z_v \end{vmatrix},\\
&\quad \left. f_v \begin{vmatrix} x_w & x_s \\ y_s & y_w \end{vmatrix} - f_w \begin{vmatrix} x_v & x_s \\ y_s & y_v \end{vmatrix} + f_s \begin{vmatrix} x_v & x_w \\ y_w & y_v \end{vmatrix} \right)\\
&= f_v \left(\begin{vmatrix} y_w & y_s \\ z_s & z_w \end{vmatrix}, - \begin{vmatrix} x_w & x_s \\ z_s & z_w \end{vmatrix}, \begin{vmatrix} x_w & x_s \\ y_s & y_w \end{vmatrix} \right)\\
&\quad -f_w \left(\begin{vmatrix} y_v & y_s \\ z_s & z_v \end{vmatrix}, - \begin{vmatrix} x_v & x_s \\ z_s & z_v \end{vmatrix}, \begin{vmatrix} x_v & x_s \\ y_s & y_v \end{vmatrix} \right)\\
&\quad +f_s \left(\begin{vmatrix} y_v & y_w \\ z_w & z_v \end{vmatrix}, - \begin{vmatrix} x_v & x_w \\ z_w & z_v \end{vmatrix}, \begin{vmatrix} x_v & x_w \\ y_w & y_v \end{vmatrix} \right)\\
&= (\mathbf{W}_v \times \mathbf{W}_s) f_v - (\mathbf{W}_v \times \mathbf{W}_s) f_w + (\mathbf{W}_v \times \mathbf{W}_w) f_s.
\end{aligned}
$$

It follows from this that

$$
\nabla f = \frac{1}{D}\big[(\mathbf{W}_v \times \mathbf{W}_s) f_v - (\mathbf{W}_v \times \mathbf{W}_s) f_w\big] + \mathbf{W}_s f_s. \tag{A.21}
$$

A.5 Divergence

Let $\mathbf{V} = (a,\ b,\ c)$. Then following the above

$$
Da_x = \begin{vmatrix} a_v & y_v & z_v \\ a_w & y_w & z_w \\ a_s & y_s & z_s \end{vmatrix}, \quad Db_x = \begin{vmatrix} b_v & y_v & z_v \\ b_w & y_w & z_w \\ b_s & y_s & z_s \end{vmatrix}, \quad Dc_x = \begin{vmatrix} c_v & y_v & z_v \\ c_w & y_w & z_w \\ c_s & y_s & z_s \end{vmatrix}
$$

$$
Da_y = \begin{vmatrix} x_v & a_v & z_v \\ x_w & a_w & z_w \\ x_s & a_s & z_s \end{vmatrix}, \quad Db_y = \begin{vmatrix} x_v & b_v & z_v \\ x_w & b_w & z_w \\ x_s & b_s & z_s \end{vmatrix}, \quad Dc_y = \begin{vmatrix} x_v & c_v & z_v \\ x_w & c_w & z_w \\ x_s & c_s & z_s \end{vmatrix}
$$

$$
Da_z = \begin{vmatrix} x_v & y_v & a_v \\ x_w & y_w & a_w \\ x_s & y_s & a_s \end{vmatrix}, \quad Db_z = \begin{vmatrix} x_v & y_v & b_v \\ x_w & y_w & b_w \\ x_s & y_s & b_s \end{vmatrix}, \quad Dc_z = \begin{vmatrix} x_v & y_v & c_v \\ x_w & y_w & c_w \\ x_s & y_s & c_s \end{vmatrix}.
$$

Then

$$
\begin{aligned}
D\,\nabla\cdot\mathbf{V} &= D(a_x+b_y+c_v)\\
&= \begin{vmatrix} a_v & y_v & z_v\\ a_w & y_w & z_w\\ a_s & y_s & z_s\end{vmatrix} + \begin{vmatrix} x_v & b_v & z_v\\ x_w & b_w & z_w\\ x_s & b_s & z_s\end{vmatrix} + \begin{vmatrix} x_v & y_v & c_v\\ x_w & y_w & c_w\\ x_s & y_s & c_s\end{vmatrix}\\
&= \begin{vmatrix} a_v & b_v & c_v\\ x_w & y_w & z_w\\ x_s & y_s & z_s\end{vmatrix} + \begin{vmatrix} x_v & y_v & z_v\\ a_w & b_w & c_w\\ x_s & y_s & z_s\end{vmatrix} + \begin{vmatrix} x_v & y_v & z_v\\ x_w & y_w & z_w\\ a_s & b_s & c_s\end{vmatrix}\\
&= \mathbf{V}_v\cdot(\mathbf{W}_w\times\mathbf{W}_s) - \mathbf{V}_w\cdot(\mathbf{W}_v\times\mathbf{W}_s) + \mathbf{V}_s\cdot(\mathbf{W}_v\times\mathbf{W}_w)\,.
\end{aligned}
$$

It follows that the divergence of $\mathbf{V}$ is given by

$$
\nabla\cdot\mathbf{V} = \frac{1}{D}\Big[\mathbf{V}_v\cdot(\mathbf{W}_w\times\mathbf{W}_s) - \mathbf{V}_w\cdot(\mathbf{W}_v\times\mathbf{W}_s)\Big] + \mathbf{W}_s\cdot\mathbf{V}_s. \tag{A.22}
$$

A.6 Curl

The curl of the vector $\mathbf{V}$ is defined as follows:

$$
\nabla\times\mathbf{V} = (c_v - b_z,\ -c_x + a_z,\ b_x - a_y).
$$

$$
\begin{aligned}
D\,(\nabla\times\mathbf{V}) = \Bigg(& \begin{vmatrix} x_v & c_v & z_v\\ x_w & c_w & z_w\\ x_s & c_s & z_s\end{vmatrix} - \begin{vmatrix} x_v & y_v & b_v\\ x_w & y_w & b_w\\ x_s & y_s & z_s\end{vmatrix}\\
&, -\begin{vmatrix} c_v & y_v & z_v\\ c_w & y_w & z_w\\ c_s & y_s & z_s\end{vmatrix} + \begin{vmatrix} x_v & y_v & a_v\\ x_w & y_w & a_w\\ x_s & y_s & a_s\end{vmatrix}\\
&, \begin{vmatrix} b_v & y_v & z_v\\ b_w & y_w & z_w\\ b_s & y_s & z_s\end{vmatrix} - \begin{vmatrix} x_v & a_v & z_v\\ x_w & a_w & z_w\\ x_s & a_s & z_s\end{vmatrix}\Bigg).
\end{aligned}
$$

This reduces to

$$
\begin{aligned}
&D\,(\nabla \times \mathbf{V}) \\
&= \Big(b_v(x_s y_w - x_w y_s) + c_v(x_s z_w - x_w z_s), \\
&\qquad -c_v(y_w z_s - y_s z_w) + a_v(x_w y_s - x_s y_w), \\
&\qquad\qquad b_v(y_w z_s - y_s z_w) + a_v(x_w z_s - x_s z_w)\Big) \\
&+\Big(c_w(x_v z_s - x_s z_v) + b_w(x_v y_s - x_s y_v), \\
&\qquad c_w(y_v z_s - y_s z_v) - a_w(x_v y_s - x_s y_v), \\
&\qquad\qquad -b_w(y_v z_s - y_s z_v) - a_w(x_v z_s - x_s z_v)\Big) \\
&+\Big(- c_s(x_v z_w - x_w z_v) - b_s(x_v y_w - x_w y_v), \\
&\qquad -c_s(y_v z_w - y_w z_v) + a_s(x_v y_w - x_w y_v), \\
&\qquad\qquad b_s(y_v z_w - y_w z_v) + a_s(x_v z_w - x_w z_v)\Big) \\
&= \Big(- (a_v x_s + b_v y_s + c_v z_s)x_w + (a_v x_w + b_v y_w + c_v z_w)x_s, \\
&\qquad -(a_v x_s + b_v y_s + c_v z_s)y_w + (a_v x_w + b_v y_w + c_v z_w)y_s, \\
&\qquad\qquad -(a_v x_s + b_v y_s + c_v z_s)z_w + (a_v x_w + b_v y_w + c_v z_w)z_s\Big) \\
&+\Big((a_w x_s + b_w y_s + c_w z_s)x_v - (a_w x_v + b_w y_v + c_w z_v)x_s, \\
&\qquad (a_w x_s + b_w y_s + c_w z_s)y_v - (a_w x_v + b_w y_v + c_w z_v)y_s, \\
&\qquad\qquad (a_w x_s + b_w y_s + c_w z_s)z_v - (a_w x_v + b_w y_v + c_w z_v)z_s\Big) \\
&+\Big(- (a_s x_w + b_s y_w + c_s z_w)x_v + (a_s x_v + b_s y_v + c_s z_v)x_w, \\
&\qquad -(a_s x_w + b_s y_w + c_s z_w)y_v + (a_s x_v + b_s y_v + c_s z_v)y_w, \\
&\qquad\qquad -(a_s x_w + b_s y_w + c_s z_w)z_v + (a_s x_v + b_s y_v + c_s z_v)z_w\Big).
\end{aligned}
$$

Now switching form a scalar to a vector notation, we get

$$
\begin{aligned}
D\,(\nabla \times \mathbf{V}) &= \Big(- (\mathbf{V}_v \cdot \mathbf{W}_s)\mathbf{W}_w + (\mathbf{V}_v \cdot \mathbf{W}_w)\mathbf{W}_s\Big) \\
&\quad +\Big((\mathbf{V}_w \cdot \mathbf{W}_s)\mathbf{W}_v - (\mathbf{V}_w \cdot \mathbf{W}_v)\mathbf{W}_s\Big) \\
&\qquad +\Big(- (\mathbf{V}_s \cdot \mathbf{W}_w)\mathbf{W}_v + (\mathbf{V}_s \cdot \mathbf{W}_v)\mathbf{W}_w\Big) \\
&= (\mathbf{W}_w \times \mathbf{W}_s) \times \mathbf{V}_v - (\mathbf{W}_v \times \mathbf{W}_s) \times \mathbf{V}_w + (\mathbf{W}_v \times \mathbf{W}_w) \times \mathbf{V}_s.
\end{aligned}
$$

The final result for the curl of **V** is then

$$
\nabla \times \mathbf{V} = \frac{1}{D}\Big[(\mathbf{W}_w \times \mathbf{W}_s) \times \mathbf{V}_v - (\mathbf{W}_v \times \mathbf{W}_s) \times \mathbf{V}_w\Big] + \mathbf{W}_s \times \mathbf{V}_s. \tag{A.23}
$$

A.7 Lagrangian

The assertion is that the Lagrangian of the scalar function $Q(v,\ w,\ s)$ is given by

$$\begin{aligned}\nabla^2 Q &= \nabla\cdot(\nabla Q)\\ &= \frac{1}{D}\Big\{\mathbf{W}_w\cdot\frac{\partial}{\partial v}\left(\frac{Q_v\mathbf{W}_w - Q_v\mathbf{W}_v}{D}\right)\\ &\quad -\mathbf{W}_v\cdot\frac{\partial}{\partial w}\left(\frac{Q_v\mathbf{W}_w - Q_v\mathbf{W}_v}{D}\right) + D_s Q_s\Big\} + Q_{ss}.\end{aligned} \tag{A.24}$$

To show this, note that

$$\begin{aligned}Q_v\mathbf{W}_w - Q_v\mathbf{W}_v &= (\nabla Q\cdot\mathbf{W}_v)\mathbf{W}_w - (\nabla Q\cdot\mathbf{W}_w)\mathbf{W}_v\\ &= -\nabla Q\times(\mathbf{W}_v\times\mathbf{W}_w) = D(\mathbf{W}_s\times\nabla Q).\end{aligned} \tag{A.25}$$

Inserting this into the right member of Eq. A.24 yields,

$$\begin{aligned}&\frac{1}{D}\Big\{\mathbf{W}_w\cdot\frac{\partial}{\partial v}(\mathbf{W}_s\times\nabla Q) - \mathbf{W}_v\cdot\frac{\partial}{\partial w}(\mathbf{W}_s\times\nabla Q) + D_s Q_s\Big\} + Q_{ss}\\ &= \frac{1}{D}\Big\{\mathbf{W}_w\cdot[(\mathbf{W}_{vs}\times\nabla Q) + \mathbf{W}_s\times(\nabla Q)_v]\\ &\quad -\mathbf{W}_v\cdot[(\mathbf{W}_{ws}\times\nabla Q) + \mathbf{W}_s\times(\nabla Q)_w] + D_s Q_s\Big\} + Q_{ss}\\ &= \frac{1}{D}\Big\{-(\mathbf{W}_v\times\mathbf{W}_w)_s\cdot\nabla Q\\ &\quad +[(\mathbf{W}_w\times\mathbf{W}_s)\cdot(\nabla Q)_v - (\mathbf{W}_v\times\mathbf{W}_s)\cdot(\nabla Q)_w] + D_s Q_s\Big\} + Q_{ss}\\ &= \frac{1}{D}\Big\{-D_s(\mathbf{W}_s\cdot\nabla Q) + D[\nabla\cdot(\nabla Q) - (\nabla Q)_s\mathbf{W}_s] + D_s Q_s\Big\} + Q_{ss}\\ &= \nabla\cdot(\nabla Q) = \nabla^2 Q,\end{aligned}$$

which was to be shown.

A.8 Directional Derivative

Let $\mathbf{A} = (a,\ b,\ c)$ and let $\mathbf{P}$ be any vector. Then

$$\begin{aligned}(\mathbf{P}\cdot\nabla)\mathbf{A} &= \Big((\mathbf{P}\cdot\nabla a),\ (\mathbf{P}\cdot\nabla b),\ (\mathbf{P}\cdot\nabla c)\Big)\\ &= \frac{1}{D}\Big(\mathbf{P}\cdot(\mathbf{W}_w\times\mathbf{W}_s)a_v - \mathbf{P}\cdot(\mathbf{W}_v\times\mathbf{W}_s)a_w + D(\mathbf{P}\cdot\mathbf{W}_s)a_s,\\ &\quad \mathbf{P}\cdot(\mathbf{W}_w\times\mathbf{W}_s)b_v - \mathbf{P}\cdot(\mathbf{W}_v\times\mathbf{W}_s)b_w + D(\mathbf{P}\cdot\mathbf{W}_s)b_s,\\ &\quad \mathbf{P}\cdot(\mathbf{W}_w\times\mathbf{W}_s)c_v - \mathbf{P}\cdot(\mathbf{W}_v\times\mathbf{W}_s)c_w + D(\mathbf{P}\cdot\mathbf{W}_s)c_s\Big)\\ &= \frac{1}{D}\Big[\mathbf{P}\cdot(\mathbf{W}_w\times\mathbf{W}_s)\mathbf{A}_v - \mathbf{P}\cdot(\mathbf{W}_v\times\mathbf{W}_s)\mathbf{A}_w + D(\mathbf{P}\cdot\mathbf{W}_s)\mathbf{A}_s\Big].\end{aligned}$$

From this we get

$$(\mathbf{P}\cdot\nabla)\mathbf{A} = \frac{1}{D}\Big[(\mathbf{W}_s\times\mathbf{P})\cdot\mathbf{W}_w\,\mathbf{A}_v - (\mathbf{W}_s\times\mathbf{P})\cdot\mathbf{W}_v\,\mathbf{A}_w\Big] + (\mathbf{W}_s\cdot\mathbf{P})\,\mathbf{A}_s. \tag{A.26}$$

A.9 Operations on W and its Derivatives

Now we are able to apply these results to the vector **W** and its derivatives. Here we use Eq. A.22 to get

$$\begin{aligned}\nabla\cdot\mathbf{W} &= \frac{1}{D}\left[(\mathbf{W}_w\times\mathbf{W}_s)\cdot\mathbf{W}_v - (\mathbf{W}_v\times\mathbf{W}_s)\cdot\mathbf{W}_w\right] + \mathbf{W}_s^2\\ &= \frac{1}{D}\left[2(\mathbf{W}_v\times\mathbf{W}_w)\cdot\mathbf{W}_s\right] + 1\\ &= 3.\end{aligned} \tag{A.27}$$

The divergences of the derivatives are done in the same way,

$$\begin{aligned}\nabla\cdot\mathbf{W}_v &= \frac{1}{D}\left[(\mathbf{W}_w\times\mathbf{W}_s)\cdot\mathbf{W}_{vv} - (\mathbf{W}_v\times\mathbf{W}_s)\cdot\mathbf{W}_v w\right] + \mathbf{W}_s\cdot\mathbf{W}_{vs}\\ &= \frac{1}{D}\left[(\mathbf{W}_{vv}\times\mathbf{W}_w)\cdot\mathbf{W}_s + (\mathbf{W}_v\times\mathbf{W}_{ww})\cdot\mathbf{W}_s\right]\\ &= \frac{1}{D}\left[(\mathbf{W}_v\times\mathbf{W}_w)_v\cdot\mathbf{W}_s\right]\\ &= \frac{D_v}{D}.\end{aligned} \tag{A.28}$$

Here we have made use of the fact that $\mathbf{W}_v\cdot\mathbf{W}_s = 0$ and that $\mathbf{W}_s = (1/D)(\mathbf{W}_v\times\mathbf{W}_w)$. In exactly the same way we can get

$$\nabla\cdot\mathbf{W}_w = \frac{D_w}{D}. \tag{A.29}$$

Now for the curls. First we will need the first and second fundamental quantities defined as

$$\begin{aligned}E &= \mathbf{W}_v^2 & F &= \mathbf{W}_v\cdot\mathbf{W}_w & G &= \mathbf{W}_w^2\\ L &= \mathbf{W}_s\cdot\mathbf{W}_{vv} & M &= \mathbf{W}_s\cdot\mathbf{W}_{vw} & N &= \mathbf{W}_s\cdot\mathbf{W}_{ww}.\end{aligned} \tag{A.30}$$

The derivatives of the first fundamental quantities are as follows:

$$\begin{aligned}E_s &= 2\mathbf{W}_v\cdot\mathbf{W}_{vs}, & F_s &= \mathbf{W}_{vs}\cdot\mathbf{W}_w + \mathbf{W}_v\cdot\mathbf{W}_{ws}, & G_s &= 2\mathbf{W}_w\cdot\mathbf{W}_{ws}\\ E_v &= 2\mathbf{W}_v\cdot\mathbf{W}_{vv}, & F_v &= \mathbf{W}_{vv}\cdot\mathbf{W}_w + \mathbf{W}_v\cdot\mathbf{W}_{ws}, & G_v &= 2\mathbf{W}_w\cdot\mathbf{W}_{wv}\\ E_w &= 2\mathbf{W}_v\cdot\mathbf{W}_{vw}, & F_w &= \mathbf{W}_{vw}\cdot\mathbf{W}_w + \mathbf{W}_v\cdot\mathbf{W}_{ws}, & G_w &= 2\mathbf{W}_w\cdot\mathbf{W}_{ww},\end{aligned} \tag{A.31}$$

from which we can see that

$$L = -E_s/2 \qquad M = -F_s/2 \qquad N = -G_s/2. \tag{A.32}$$

In addition note that, since $\mathbf{W}_s \cdot \mathbf{W}_v$ and $\mathbf{W}_s \cdot \mathbf{W}_w$ both equal zero,

$$\mathbf{W}_{vs} \cdot \mathbf{W}_w = \mathbf{W}_{ws} \cdot \mathbf{W}_v = -\mathbf{W}_s \cdot \mathbf{W}_{vw} \,. \tag{A.33}$$

Now turn to the calculations of the curls using Eq. A.23. First we show that the curl of $\mathbf{W}$ is zero.

$$\begin{aligned}
\nabla \times \mathbf{W} &= \frac{1}{D}\big[(\mathbf{W}_w \times \mathbf{W}_s) \times \mathbf{W}_v - (\mathbf{W}_v \times \mathbf{W}_s) \times \mathbf{W}_w\big] \\
&= -\frac{1}{D}\big[(\mathbf{W}_v \cdot \mathbf{W}_s)\mathbf{W}_w - (\mathbf{W}_v \cdot \mathbf{W}_w)\mathbf{W}_s \\
&\qquad -(\mathbf{W}_w \cdot \mathbf{W}_s)\mathbf{W}_v + (\mathbf{W}_w \cdot \mathbf{W}_v)\mathbf{W}_s\big] \\
&= 0,
\end{aligned} \tag{A.34}$$

from Eq. A.33. In exactly the same way we can show that $\nabla \times \mathbf{W}_s = 0$. Next we do $\mathbf{W}_v$, which is

$$\begin{aligned}
\nabla \times \mathbf{W}_v &= \frac{1}{D}\big[(\mathbf{W}_w \times \mathbf{W}_s) \times \mathbf{W}_{vv} - (\mathbf{W}_v \times \mathbf{W}_s) \times \mathbf{W}_{ww}\big] + \mathbf{W}_s \times \mathbf{W}_{vs} \\
&= -\frac{1}{D}\big[(\mathbf{W}_{vv} \cdot \mathbf{W}_s)\mathbf{W}_w - (\mathbf{W}_{vv} \cdot \mathbf{W}_w)\mathbf{W}_s \\
&\qquad -(\mathbf{W}_{vw} \cdot \mathbf{W}_s)\mathbf{W}_v + (\mathbf{W}_{vw} \cdot \mathbf{W}_v)\mathbf{W}_s\big] + \mathbf{W}_s \times \mathbf{W}_{vs} \\
&= \frac{1}{2D}\big[E_s\mathbf{W}_w - F_s\mathbf{W}_v + 2(E_w - F_v)\mathbf{W}_s\big] + \mathbf{W}_s \times \mathbf{W}_{vs}.
\end{aligned} \tag{A.35}$$

The calculation of curl $\mathbf{W}_w$ proceeds in the same way and is

$$\nabla \times \mathbf{W}_w = \frac{1}{2D}\big[F_s\mathbf{W}_w - G_s\mathbf{W}_v + 2(F_w - G_v)\mathbf{W}_s\big] + \mathbf{W}_s \times \mathbf{W}_{ws}. \tag{A.36}$$

A.10 An Additional Lemma

We sometimes encounter expressions of the type

$$\mathbf{W}_w \cdot \mathbf{V}_v - \mathbf{W}_v \cdot \mathbf{V}_w. \tag{A.37}$$

Expand this in the following way. Let $\mathbf{V} = (\xi,\ \eta,\ \zeta)$ and let $\mathbf{W} = (x,\ y,\ z)$ so that Eq. A.37 becomes

$$
\begin{aligned}
&(x_w,\ y_w,\ z_w)\cdot(\xi_v,\ \eta_v,\ \zeta_v) - (x_v,\ y_v,\ z_v)\cdot(\xi_w,\ \eta_w,\ \zeta_w)\\
&= (x_w\xi_v + y_w\eta_v + z_w\zeta_v) - (x_v\xi_w + y_v\eta_w + z_v\zeta_w)\\
&= x_w(\xi_x x_v + \xi_y y_v + \xi_z z_v) + y_w(\eta_x x_v + \eta_y y_v + \eta_z z_v) + z_w(\zeta_x x_v + \zeta_y y_v + \zeta_z z_v)\\
&\quad - x_v(\xi_x x_w + \xi_y y_w + \xi_z z_w) + y_v(\eta_x x_w + \eta_y y_w + \eta_z z_w) + z_v(\zeta_x x_w + \zeta_y y_w + zeta_z z_w)\\
&= (y_v z_w - z_v y_w)(\zeta_y - \eta_z) - (x_v z_w - z_v x_w)(\xi_z - \zeta_x) + (x_v y_w - y_v x_w)(\eta_x - \xi_y)\\
&= (\mathbf{W}_v \times \mathbf{W}_w)\cdot(\nabla \times \mathbf{V}) = D\mathbf{W}_s \cdot (\nabla \times \mathbf{V}).
\end{aligned}
\tag{A.38}
$$

B Bibliography

Altrichter and Schäfer 1956: O. Altrichter and G. Schäfer, "Herleitung der Gullstrandschen Grundgleichungen für Schiefe Strahlenbüschel aus den Hauptkrümmungen der Wellenfläche" *Optik(Stuttgart)* **13**, 241–253.

Avendaño-Alejo and Stavroudis 2002: M. Avendaño-Alejo and O.N. Stavroudis, "Huygens principle and rays in uniaxial anisotropic media. II. Crystal axis orientation arbitrary." *J. Opt. Soc. Am. A* **19**, 1674–1679.

Avendaño-Alejo et al. 2004: M. Avendaño-Alejo, J. Orozco-Arellanes and O.N. Stavroudis, "Spherical aberrations analysis of an uniaxial Cartesian oval." *SPIE* **5622**, 1164–1168.

Blaschke 1945: Wilhelm Blaschke, *Vorlesungen über Differential Geometrie.* Dover Publications, New York.

Bliss 1925: Gilbert A. Bliss, *Calculus of Variations.* The Mathematical Association of America, Carus Mathematical Monograph #19, Open Court Publishing Co, LaSalle, Indiana.

Bliss 1946: Gilbert A. Bliss, *Lectures on the Calculus of Variations.* The University of Chicago Press.

Bolza 1961: Oskar Bolza, *Lectures on the Calculus of Variations.* Dover Publications, New York. [1904: University of Chicago Press.]

Born and Wolf 1999: Max Born and Emil Wolf, *Principles of Optics.* 7th Edition. Cambridge University Press.

Buchdahl 1970: H. A. Buchdahl, *An Introduction to Hamiltonian Optics.* Cambridge University Press.

Buchdahl 1972: H. A. Buchdahl, "Systems without Symmetries: Foundations of a Theory of Lagrangian Aberration Coeficients." *J. Opt. Soc. Am. A* **62** 1314–1324.

Buchroeder 1970: R. A. Buchroeder," Tilted Component Telescopes. Part 1: Theory." *Appl. Opt.* **9**, 2169–2171.

Cagnet et al 1962: M. Cagnet, M. Françon and J. C. Thrier, *The Atlas of Optical Phenomena.* Prentice Hall, Englewood Cliffs, N. J.

Carathéodory 1989: C. Carathéodory, *Calculus of Variations and Partial Differential Equations of the First Order.* Tr. Robert B. Dean, Tr. Editor Julius J. Brandstatter. AMS

The Mathematics of Geometrical and Physical Optics: The k-function and its Ramifications. O.N. Stavroudis

ISBN: 3-527-40448-1

Chelsea Publishing, American Mathematicial Society, Providence, RI. [*Variationsrechnung und Partielle Differentialgleichnungen erster Ordnung.* 1935: B. G. Teubner, Berlin.]

Cohen 1933: Abraham Cohen, *An Elementary Treatise on Differential Equations.* Second Edition. D. C. Heath & Co.

Coddington 1829–1830: H. Coddington, *A System of Optics* in two volumes. Limkin and Marshal, London.

Courant and Robbins 1996: Richard Courant and Herbert Robbins, *What is Mathematics?* Revised by Ian Stewart, Oxford University Press, New York.

Clegg 1968: John C. Clegg, *Calculus of Variations.* John Wiley & Sons, New York.

Conway and Synge 1931: A. W. Conway and J. L. Synge, *The Mathematical Papers of Sir William Rowan Hamilton.* Vol. I. Geometrical Optics. Cambridge University Press.

Dickson 1914: Leonard Eugene Dickson, *Elementary Theory of Equations.* John Wiley & Sons, New York.

Dickson 1939: Leonard Eugene Dickson, *New First Course in the Theory of Equations.* John Wiley & Sons, New York.

Do Carmo 1976: Manfredo P. Do Carmo, *Differential Geometry of Curves and Surfaces.* Prentice-Hall, Englewood Cliffs, New Jersey.

Flores-Hernandez and Stavroudis 1996: R. Flores-Hernandez and O.N. Stavroudis, "Real time manipulations of the Delano diagram as a linear tool for optical system optimization", *Optik* **4**, 141–147.

Forsyth 1959: Andrew Russell Forsyth, *Theory of Differential Equations.* Six volumes bound as three. Dover Publications, New York. [A reprint of Cambridge University Press, 1906.]

Forsyth 1996: Andrew Russell Forsyth, *A Treatise on Differential Equations.* Sixth Edition. Dover Publications, Mineola, New York.

Garabedian 1998: P. R. Garabedian, *Partial Differential Equations.* AMS Chelsea Publishing, Providence, Rhode Island.

Gelles 1974: R. Gelles, "Unobscured Aperture Stigmatic Telescopes." *Opt. Eng.* **13**, 534–538.

Gelles 1975: R. Gelles, "Unobscured Aperture Two-Mirror Systems." *J. Opt. Soc. Am. A* **65**, 1141–1143.

Goursat 1904: Édouard Goursat, *A Course in Mathematical Analysis.* Translated ed by Earle Raymond Heddrick. Ginn & Co. Boston.

Goursat 1945: Édouard Goursat, *Differential Equations.* Translated ed by Earle Raymond Heddrick and Otto Dunkel. Ginn & Co. Boston.

Gugenheimer 1977: Heinrich W. Gugenheimer, *Differential Geometry.* Dover Publications, New York.

Gullstrand 1906: A. Gullstrand, "Die reelle optische Abbildung." *Sv. Vetensk. Handl.* **41**, 1–119.

Herzberger 1935: Max Herzberger, "On the fundamental optical invariant, the optical tetrality principle, and on the new development of Gauss optics based on this law." *J. Opt. Soc. Am.* **25**, 197–204

Herzberger 1936: Max Herzberger, "A new theory of optical image formation." *J. Opt. Soc. Am.* **26**, 295–304.

Herzberger 1937: Max Herzberger, "Optics in the large." *J. Opt. Soc. Am.* **27**, 202–206.

Herzberger 1939: Max Herzberger, "Normal systems with two caustic lines." *J. Opt. Soc. Am.* **29**, 392–394.

Herzberger 1940: Max Herzberger, "Normal systems with two caustic lines." *J. Opt. Soc. Am.* **30**, 307–308.

Herzberger 1943a: Max Herzberger, "Direct methods in geometrical optics." *Trans. Am. Math. Soc.* **53**, 218–229.

Herzberger 1943b: Max Herzberger, "A direct image error theory." *Q. Appl. Math.* **1**, 69–77.

Herzberger 1946: Max Herzberger, "The Limitation of Optical Image Formation, Points and Diapoints." *Ann. N. Y. Acad. Sci.* **48**, 7–13.

Herzberger 1954a: Max Herzberger, "Some remarks on ray tracing." *J. Opt. Soc. Am.* **41**, 805–807.

Herzberger 1954b: Max Herzberger, "Image errors and diapoint errors." In *Studies in Mathematics and Mechanics presented to Richard von Mises by Friends, Colleagues and Pupils.* Academic Press, New York. pp. 30–35.

Herzberger 1958: Max Herzberger, *Modern Geometrical Optics.* John Wiley & Sons (Interscience), New York.

Itô 1993: Kiyosi Itô, Editor. *Encyclopedic Dictionary of Mathematics.* Second Edition. MIT Press, Cambridge MA and London.

Joos 1941: G. Joos, *Theoretical Physics.* Tr. I. M. Freeman, Blackie & Son, London.

Kline and Kay 1965: M. Kline and I. W. Kay, *Electromagnetic Theory and Geometrical Optics.* Interscience/John Wiley & Sons.

Kemmer 1977: N. Kemmer, *Vector Analysis. A physicist's guide to the mathematics of fields in three dimensions.* Cambridge University Press.

Kneisly 1964: J. A. Kneisly III, "Local curvatures of wavefronts in an optical system." *J. Opt. Soc. Am.* **54**, 229–235.

Korn and Korn 1968: G. A. Korn and T. M. Korn, *Mathematical Handbook for Scientists and Engineers.* Second Edition. McGraw-Hill, New York.

Korsch 1991: Dietrich Korsch, *Reflective Optics*. Academic Press, San Diego.

Kutter 1953: Anton Kutter, *Der Schiefspiegler*. Biberacher Verlagsdruckerei, Biberach an der RiS, Germany.

Kutter 1969: A. Kutter, *Sky and Telescope 38* 418 (1969). Cited in Buchroeder 1970.

Luneburg 1964: R. K. Luneburg, *Mathematical Theory of Optics*. University of California Press, Berkeley. Reprint of notes given for a course at Brown University, 1944.

Malacara and Thompson 2001: Daniel Malacara and Brian J. Thompson, *Handbook of Optical Engineering*. Marcel Dekker, Inc., New York.

Marion and Heald 1980: J. B. Marion and M. A. Heald, *Classical ElectroMagnetic Radiation*. 2nd Ed. Academic Press, New York.

Maxwell 1952: James C. Maxwell, *Scientific Papers*. W. D. Niven, Editor. Dover Publications, New York.

O'Connell 1998: James O'Connell, "Review of *Maxwell on the Electromagnetic Field: A Guided Study*." *Am. J. Phys.* **66**, 92–93.

Rektorys 1969: Karel Rektorys, *Survay of Applicable Mathematics*. The M. I. T. Press, Cambridge, Massachusetts.

Sabra 1967: A. I. Sabra, *Theories of Light. From Descartes to Newton*. Oldbourne, London.

Sands 1972: P. J. Sands, "Aberration Coeficients of Plane Symmetric Systems." *J. Opt. Soc. Am. A* **62**, 1211–1220.

Sands 1973a: P. J. Sands, "Aberration Coeficients of Double-Plane[a] Symmetric Systems." *J. Opt. Soc. Am. A* **63**, 425–430.

Sands 1973b: P. J. Sands, "Thin Double-Plane Symmetric Systems." *J. Opt. Soc. Am. A* **63**, 431–434.

Smith 1959: David Eugene Smith, *A Source Book in Mathematics*. Dover Publications, New York. [1929: McGraw-Hill Book Co.]

Stavroudis and Sutton 1965: O. N. Stavroudis and L. E. Sutton, *Spot Diagrams for the Prediction of Lens Performance from Design Data*. NBS Monograph No. 93 U. S. Government Printing Office. Washington, DC.

Stavroudis 1967: O. N. Stavroudis, "Resolving power predictions from lens design data." *Appl. Opt.* **6** 129–135.

Stavroudis 1972a: O. N. Stavroudis, *The Optics of Rays, Wavefronts and Caustics*. Academic Press, New York.

Stavroudis 1972b: O. N. Stavroudis, "Some consequences of Herzberger's fundamental optical invarient." *J. Opt. Soc. Am.* **62**, 59–63.

Stavroudis 1976: O. N. Stavroudis, "A simpler derivation of the formulas for generalized ray tracing." *J. Opt. Soc. Am.* **66**, 1330–1333.

Stavroudis and Fronczek 1976: O. N. Stavroudis and R. C. Fronczek, "Caustic surfaces and the structure of the geometric image." *J. Opt. Soc. Am.* **66**, 795–800.

Stavroudis et al. 1978: O. N. Stavroudis, R. Fronczek and R.-S. Chang, "The geometry of the half symmetric image." *J. Opt. Soc. Am.* **68**, 739–742.

Stavroudis and Fronczek 1978: O. N. Stavroudis and R. Fronczek, "Generalized ray tracing and the caustic surface." *Optics and Laser Technology* **10**, 185–191.

Stavroudis and Ames 1992: O. N. Stavroudis and A. J. Ames, "Confocal Prolate Spheroids in an Off-Axis System." *J. Opt. Soc. Am. A* **9**, 2083–2088 (1992).

Stavroudis 1995a: O. N. Stavroudis, "The k-function in geometrical optics and its relationship to the archetypical wave front and the caustic surface." *J. Opt. Soc. Am. A* **12**, 1010–1016.

Stavroudis 1995b: O. N. Stavroudis, "Refraction of the k-function at spherical surfaces." *J. Opt. Soc. Am. A* **12**, 1805–1811.

Stavroudis 1996: O. N. Stavroudis,"The caustic as an expression of the image errors of a lens." In: *OSA Proceedings of the International Optical Design Conference, 1994*, Vol. 22, Ed. G. W. Forbes, pp. 87–91.

Stavroudis and Flores-Hernandez 1998: O. N. Stavroudis and R. Flores-Hernandez, "Structure of the modern schiefspiegler. I. Pseudo axis, magnification, and distortion", *J. Opt. Soc. Am. A* **15**, 437–442.

Stavroudis 2001: "Basic Ray Optics." In: *Handbook of Optical Engineering* Daniel Malacara and Brian J. Thompson, Eds. Marcel Decker, New York and Basel. pp. 1–38.

Stoker 1969: J. J. Stoker, *Differential Geometry*. Wiley–Interscience, New York.

Stone and Forbes 1992a: B. D. Stone and G. W. Forbes, "Foundations of First-Order Layout for Asymmetric Systems: An Application of Hamilton's Methods." *J. Opt. Soc. Am. A* **9**, 96–109.

Stone and Forbes 1992b: B. D. Stone and G. W. Forbes, "First Order Layout of Asymmetric Systems Composed of Three Spherical Mirrors." *J. Opt. Soc. Am. A* **9**, 110–120.

Stone and Forbes 1992c: B. D. Stone and G. W. Forbes, "Characterization of First-Order Optical Properties for Asymmetric Systems." *J. Opt. Soc. Am. A* **9**, 478–489.

Stone and Forbes 1992d: B. D. Stone and G. W. Forbes,"Forms of the Characteristic Function for Asymmetric Systems that Form Sharp Images to the First Order." *J. Opt. Soc. Am. A* **9**, 820–831.

Stone and Forbes 1992e: B. D. Stone and G. W. Forbes, "First-Order Layout of Asymmetric Systems: Sharp Imagery of a Single Plane Object." *J. Opt. Soc. Am. A* **9**, 832–843
.

Struik 1961: Dirk J. Struik, *Lectures on Classical Differential Geometry*. Second Edition. Addison-Wesley, Reading, MA.

Todhunter 1961: I. Todhunter, *A History of the Calculus of Variations.* Chelsia Publishing Co., New York. [A reprint of *A History of the Progress of the Calculus of Variations during the Nineteenth Century.* 1861: Cambridge]

Troutman 1983: John L. Troutman, *Variational Calculus with Elementary Convexity.* Springer-Verlag, New York.

Washer 1965: F. E. Washer, *Phot.Eng.* **52**, 213 (1966).

Welford 1974: W. T. Wellford, *Aberrations of the Symmetrical Optical System.* Academic Press, London.

Woodhouse 1964: Robert Woodhouse, *A History of the Calculus of Variations in the Eighteenth Century.* Chelsea, Bronx, NY. [A reprint of *A Treatise on Isoperimetrial Problems and the Calculus of Variations.* 1810.]

Young E. 1993: Eutiquio C. Young, *Vector and Tensor Analysis.* Second Edition. Marcel Dekker, New York.

Young L. 1969: L. C. Young, *Lectures on the Calculus of Variations and Optimal Control Theory.* W. B. Saunders Company, Philadelphia.

Index

The Mathematics of Geometrical and Physical Optics: The k-function and its Ramifications. O.N. Stavroudis

ISBN: 3-527-40448-1